ORGONE ENERGY

Previous Books by Jerome Eden

Memoir of F. A. Mesmer, 1799 (Translator)
Maxims on Animal Magnetism (Translator)
Suffer the Children

Orgone Energy

THE ANSWER TO ATOMIC SUICIDE

by

JEROME EDEN

Foreword by Richard A. Blasband, M.D.
Fellow, American College of Orgonomy

EXPOSITION PRESS NEW YORK

EXPOSITION PRESS INC.
900 So. Oyster Bay Road, Hicksville, N.Y. 11801

1st printing, June, 1972
2nd Printing, October, 1974

SBN 0-682-47477-0

This book is dedicated to my friend Elsworth F. Baker, M.D., who has done more than anyone else I know to further the legacy left by Wilhelm Reich.

DERBY SQUARE BOOK STORE
215 Essex Street
Salem, Massachusetts 01970

11/24/75

Dear Thomas:

The books you ordered:

Planet in Trouble: The U.F.O. Assault On Earth by Jerome Eden
and Orgone Energy by Jerome Eden

have arrived.

You may pick it up here at the book store at your convenience.

Sincerely,
Ted Monroe

Thomas Cadoret
Hillside,
Rowley, Ma.

Contents

Foreword

This is an important book. In a personal, immediate, readily comprehensible way, it tells of the havoc created by nuclear radiation today. The damage, as Mr. Eden so forcefully and dramatically describes, is caused not by the direct effects of radiation per se, but through its excitation of the orgone energy fields of the Earth and all the living things upon it.

Wilhelm Reich's discovery of orgone energy, an atmospheric energy which is also the life energy, is Man's greatest scientific achievement of the twentieth century. Traditional science has yet to recognize this, but it must do so in the near future if we are to avoid the chaos that will result from acute power failures and the threat of slow extinction from deleterious alterations in our life energy. There is ample evidence that both events are already taking place. The former needs no documentation, the latter is well described herein: it is the antagonism between nuclear radiation and orgone energy and the resulting production of a deadly variation of the orgone.

Jerome Eden is qualified to tell this story. Though he is not trained as a scientist, he has an immediate and fully contactful grasp of Reich's work and has had much personal experience with orgone energy and its derivatives. He has fought valiantly to protect our natural heritage against nuclear pollution, yet well understands how we must also defend our freedom against the formidable threat of Red Fascism. Life-negating, mechanistic science, the product of rigidified emotions, has forced us into a paradoxical situation where we must defend ourselves against one life-inimical force by use of another. Our only hope is in the realization of the life-positive power of orgone energy. The knowledge is there in Reich's work, if man wishes to assume responsibility for it. We may be grateful to Jerry Eden for doing his part.

RICHARD A. BLASBAND, M.D.
Fellow, American College of Orgonomy

Preface

Since the dawn of Life on the planet Earth, men have searched for the answers to the questions: Who am I? Where do I come from and where am I going? What is Life? What is Death?

The almost simultaneous discovery of atomic energy and Orgone energy brings these questions into sharp focus. Knowledge and use of the atom, without an understanding of Orgone energy, will toll the death knell for all of Life. That, in substance, is what this book is all about.

In these pages the reader will meet—perhaps for the first time—facts and concepts dealing with an all-pervasive power *infinitely greater than atomic energy*. The man who discovered this hitherto unknown energy is Wilhelm Reich, M. D. He gave his life to place on official court record the facts dealing with his discoveries and the incredible opposition placed in his path by the Enemies of Mankind.

The situation is critical; the hour is very late. It is my fervent prayer that God will give to men and women of good will the wisdom to discern good from evil, and the strength to do what must be done in this Final Fight for Life.

JEROME EDEN

Acknowledgments

The author acknowledges with thanks permission granted by the Editor of *The Journal of Orgonomy* to use material from that publication. Thanks also go to Brad Steiger for permission to quote from Chapter 3, "The Silencing of Wilhelm Reich," appearing in the book, *The Allende Letters–New UFO Breakthrough;* and to John W. Gofman, M. D., and Arthur R. Tamplin, Ph.D., for allowing me to quote from their book, *Poisoned Power.*

ORGONE ENERGY

"Only the truth became too preposterous to entertain."

WHITTAKER CHAMBERS

1 Who Wants Truth?

"And ye shall know the truth, and the truth shall make you free," quoted the disciple John nearly 2,000 years ago, just a few years after they hung a great man named Jesus Christ on a cross for speaking truth.

In our 20th century, a very great scientist and humanitarian said, "Truth is immediate contact with reality." That scientist was Dr. Wilhelm Reich. For investigating and publishing scientific truths for more than 30 years he was rewarded with a prison term in a Federal penitentiary, where he died in November 1957.

If we investigate the history of science we find that whenever a man—regardless of the century—stumbles upon some great *new* truth, and pursues it honestly and effectively, his reward is death—by outright murder, by ostracism, by loss of income, or by vicious slander that eventually breaks his spirit or his heart.

Not too long ago, I found myself in the position of attempting to bring some vital truths about atomic energy pollution to my fellow citizens of Idaho Falls. I was an active member of the community, project director of the Eastern Idaho Special Services Agency, funded by the State of Idaho's Administration on Aging and the U.S. Office of Economic Opportunity. Prior to that, I was city and military editor of the Idaho Falls *Post-Register*, Idaho's second largest newspaper. I had my own weekly radio program in which I discussed the problems of the area's senior citizens.

Some 50 miles west of Idaho Falls, out in the desert, sits the sprawling complex of the National Reactor Testing Station. Dirty brownish-yellow smoke hangs in mile-long plumes east of the NRTS. The prevailing winds are from west to east. Millions upon millions of gallons of radioactive waste materials are buried (and continue to be buried) a few scant feet below the desert floor at the NRTS site. Beneath these countless tons of lethal radioactivity runs the Snake River aquifer—one of the largest underground bodies of fresh water in the nation, water which nourishes most of the Northwest. The area has been rated by the U.S. Geological Survey Dept. as one of the seven most earthquake-prone regions in the entire United States!

As I write these lines, the city of Los Angeles is digging out from under its worst earthquake disaster in nearly four decades. What if such an earthquake had struck southeastern Idaho?

I consider the disaster potential in the Idaho Falls area far greater than the recent earthquake in Los Angeles. I considered it dangerous enough to stick my neck out in attempting to alert the local citizenry to the tremendous hazards involved.

On November 14, 1969, I placed the following display ad in the *Post-Register:*

IS RADIOACTIVITY

THREATENING YOUR LIFE?

What Can You Do

To Help Yourself

And Your Children?

Idaho Citizens for a Healthy Environment
Will Meet Saturday at 8 p.m.
At 1125 S.E. Bonneville Dr. Phone 523-4581
—All Welcome—

This two-column four-inch ad was enclosed in a heavy black border. For the past several months I had done everything I knew to alert my fellow citizens to the dangers that faced us. I had appeared on local television, had made statements which were printed in the newspaper and released to the radio. As the day for the Idaho Citizens for a Healthy Environment meeting drew near, my mute telephone seemed to become a silent mocker of my folly. "Why knock yourself out, Jerry? People don't want the truth. And if you bother them too much you might very well end up—all alone in the radioactive air—hanging on a cross!" That's what my telephone was telling me.

At the appointed date and time, the Idaho Citizens for a Healthy Environment, consisting of myself, my wife, Desiree, and our two dogs, Faith and Sergeant, waited in our living room for the doorbell to ring. Benches and chairs were set up to accommodate up to 40 people. At last the doorbell rang. In walked a middle-aged couple. The wife was interested in radiation poisoning because her father had died from radioactive contamination as a result of scientific work he had done some years before. The woman's husband was an employee of the National Reactor Testing Station. The only reason he was attending the meeting, he informed us, was that his wife wanted to come.

My wife served coffee while we sat, a bit uncomfortably, waiting for the "rest" of the group to arrive. At ten minutes to eight, the doorbell rang again. In walked three young men (one with his wife), and my wife asked them to sign our register. As they were signing the book, Desiree asked: "I'll bet all three of you work for the National Reactor Testing Station?" Rather sheepishly the three admitted their affiliation—they were health physicists.

Later we learned that the young scientists had "practically been ordered" to attend our meeting in order to "find out what was going on." In retrospect, I'm sure that these three young scientists must have had quite a laugh relating to their superiors and fellow workers the details of our "committee"—a meeting attended by me, my wife, one concerned woman, her chauffeur-husband, and two dogs.

Nevertheless, atomic pollution is not a matter for mere amusement—it is *deadly serious business.* H. G. Wells wrote, "Human history becomes more and more a race between education and catastrophe." And what is education if not a knowledge of the truth?

But what is "truth" to people who do not want it or cannot tolerate it? What is "truth" about radioactivity to a man who works for the Atomic Energy Commission?—and several thousand people in Idaho Falls work for the NRTS. When it comes to a battle between the truth and financial security, the truth falls upon deaf ears, or is sacrificed.

Thus it happened one day. The chairman of my board of directors came into my office and told me he had received several telephone calls from members of the local Community Chest advising me to "cool" my criticism of the NRTS. The advice carried an implied economic threat, inasmuch as my agency had only recently been voted a participating member of the Community Chest. It was no secret that the employees at the NRTS, as a group, contributed the largest amount of money to the annual Community Chest drive.

I informed my board chairman that all of my criticism of the NRTS was being done as a private citizen. He said he understood that fact; however, he felt obliged to deliver the message to me. A few days later a woman came to my office on some matter. She informed me that her husband worked at "the site" (NRTS) and that "a number of men out there are boiling mad at you and are threatening to ride you out of town on a rail." Shortly thereafter, at a breakfast meeting of prominent members of the community, a public relations man with the NRTS, who had worked with me on a project instituted by Idaho's Governor Don W. Samuelson, approached me and engaged me in some trivial conversation. His parting words were: "Well, I hope you're going to be a good boy now!" When I asked him what he meant by that remark, he refused to elaborate. But we both knew very well he was referring to my critical remarks about the NRTS.

Many times during the battle over atomic pollution I have asked myself: "What are you fighting for?" The answer is simple: I am fighting for *Life*—for my life, my wife's life, for the lives of millions of infants and children who know not what is killing or maiming them. I am fighting for Life in the atmosphere, in the earth, in the waters—the fish, the fowl, and the animals that are dying by the millions; the timber and wildflowers that are drooping and dying under the power of a deadly Radioactive Killer that will, if unchecked, destroy the very planet Earth which supports and succors us. And last, but far from least, I fight for the lives of the millions of unborn who will inherit a planet suffocating and strangling in waste products.

In brief, the *antagonist* forces me to Fight for Life—*all Life*—not because I want to, but because I must! My case against atomic pollution (and any other pollution that affects my life) is very simple: I maintain that *no one has the right to drive his automobile into my house and leave his engine running, without my expressed permission!* My "house" in this analogy is my body; the car exhaust is radioactive pollution.

I therefore charge the Radioactive Polluters of this planet with *criminal negligence!* This charge is made in all sincerity, and with the deepest conviction that time is running out for all of us. The charge of criminal negligence must be answered by the Radioactive Polluters of this and future generations. Negligence, at law, is the failure to act "as reasonable men" under the circumstances.

In view of the facts which will be brought out in these pages, I believe the charge of criminal negligence is absolutely valid. You, the readers—the men, women, and young people of this earth—are the only ones who can call a halt to the further devastation of our biosphere by "atomic madmen who know not what they do."

I pray to God that there is yet time enough to act.

2 *Rejected Prophecy*

Men reject their prophets and slay them.
Dostoyevsky

In the early 1960s I was working as managing editor of the *Journal American Water Works Association*, the leading scientific journal in the field of public water supply. At that time, AWWA had some 17,000 members all over the world, many of them leading experts in the area of public water supplies. AWWA typifies professional associations everywhere and in all fields. If a professor of water-quality treatment, biochemistry, or water engineering wrote a paper on his subject, he would submit it to the *JAWWA* for publication. The editorial staff acted merely in a mechanical capacity. The paper was redirected to one of several committees which read papers and either accepted or rejected them. Once a paper was accepted for publication, it underwent a number of processing steps before it appeared in the highly stylized coated-stock pages of our publication.

If, however, someone submitted a paper dealing with a subject which was, say "controversial," or outside the ken of the established committees, it had little opportunity of finding its way onto the printed page. No matter how much work a man did in some "borderline" field, his chances of having his paper published were usually equivalent to those of the proverbial snowball in hell.

On several occasions I had heated discussions with the assistant director of publications over the rejection of some "controversial" paper that had been submitted, or over the refusal even to review a new book in some *new* field of research. (When you work in an established field for a number of years, you soon develop a nose for what is "acceptable" and what is "unacceptable.") The assistant director informed me that you can always spot a "reactionary kook" by any one of the following criteria: (1) he believes that the fluoridation of public water supplies is a Communist plot to weaken or destroy America; (2) he believes that UFOs (Unidentified Flying Objects) come from outer space;

(3) he believes in water dowsing; (4) he believes in some "new and revolutionary idea" for creating rain, other than the "acceptable" method of cloud-seeding with silver iodide; (5) he believes that the use of pesticides, herbicides, DDT, and radioactivity is dangerously contaminating our planet's water supplies.

Rachel Carson's magnificent book *Silent Spring* was published during this period. She was crucified by know-it-alls who gleefully lost no chance to clobber her work. The scientific fraternity—with a few notable exceptions—seemed to draw together in a united front against this "irresponsible and exaggerating alarmist" who, they alleged, didn't really know what in blazes she was talking about. Sly thrusts were taken at her book from the sanctity of the accepted bastions of science. It is my belief that such unkind cuts contributed to her untimely death of cancer. Slander and scurrility can kill just as effectively as out-and-out poison.

Today, of course, DDT has become anathema. Many scientists have jumped on the bandwagon to beat the drums against ecological pollution. One wonders where all those voices were hiding when *Silent Spring* was being vilified.

Radiation, however, is an even more subtle and insidious can of worms. For one thing, as a form of energy, you cannot readily observe it. Its "insult" upon the human organism is often so slow and subtle that few people are aware, until it is too late, of what is actually killing them. The effects of radiation poisoning are often confused with symptoms called "the flu," or "virus X, Y, or Z." Furthermore, chronic radiation poisoning will hit each individual in his weakest spot, so that while one person will complain, "Oh, I'm just having another bout of hepatitis," another will think, "It's my old migraine headache again," or, "It's just my asthma," or, "It's merely my arthritis."

The continuous onslaught of radiation poisoning is *cumulative in the human body*. It can be compared to repeated hammer blows upon a gong with a crack in it—the "crack" is our weakest spot, be it hypertension or emphysema. If you keep hitting the gong often enough (there do not have to be massive blows, but merely *constant* blows), the crack widens. Or, if you will, chronic radiation poisoning might be compared to pressure inside a boiler. If the pressure builds up continually (as does biological "pressure" under constant low-level radioactivity), the boiler will eventually burst at its weakest spot!

On the social scene, our earth appears to be "cracking" at all its chronic strain points—and all kinds of social battles are taking place. We see social stress and tension with respect to integration, the drug scene, Protestants fighting Catholics over "Irish freedom," the Women's Liberation front fighting for "freedom for women," rebellious students fighting for "student freedom," the Poles and Czechoslovakians fighting for "workers' freedom," and adolescents fighting for anything that appears to be a likely "cause." National and international tensions are at a critical point as well, and have been for some time.

At the risk of being called a simplistic simpleton I suggest that such a fever pitch of destructive *activity* may have as its source the ever-increasing pressure which is a concomitant of *radioactivity*! (One spring in the late 1950s, I was teaching in a New York State, junior high school. Several nations had been conducting nuclear tests, and the kids were acting "hopping wild." An older teacher confessed to me: "For some crazy reasons I find myself hitting more kids this week than I have in all my years of teaching!")

Let us review some news items *published more than ten years ago.* These are some of the items that made headlines in 1955:

(UP) "H-Bomb Fallout Could Erase Life for 7000 Square Miles" . . . (AP) "Casualties in Atomic Labs: University of California's Dr. Ralph W. Chaney says atomic laboratories 'having casualties' from radiation" . . . (AP) "Senate to Probe Atomic Dust" . . . (N.Y. *Times*) "Mutations from Atomic Radiation: AEC Spokesman says 'many researchers had come to individual conclusions' " . . . (UP) "Ohio Pond Abounds in 8-Legged Frogs" . . . (N.Y. *Journal-American*) "Study Mystery 'Fallout' Covering Upstate [N.Y.] Area: Slightly Radioactive" . . . (N.Y. *Times*) "Columbia University Starting Atomic Lab."

All of the above headlines appeared in February 1955. I wonder what ever happened to that Senate "probe" into atomic dust? I firmly believe that the political invention of "the probe" is equaled only by the excellent evasive tactic called the "appointment of a committee"—one never again hears from either "the probe" or "the committee." Then (see above) the Senate probe continues to probe into atomic dust (to find God knows what? I mean, what were they supposed to *find* in that atomic dust anyway?); but while the Senate probe is probing, an AEC spokesman says, concerning the possibility of radioactive mutations, that "many researchers had come to individual conclusions." Such a say-nothing statement serves to do exactly that—*say nothing*! Meantime, back at the frog pond in Ohio, some very brash frogs are embarrassing the hell out of the nuclear know-it-alls by daring to hop around on *eight legs*!

That was in February 1955. Now, let us look at May 1955 headlines:

"Global Research in Deserts Urged" . . . "New Study of Cloud Seeding" . . . (N.Y. *Times*) "Cloud Seeding Fails as a Storm Control: Great Trout Loss Due to Lack of Rain." (Be careful there, N.Y. *Times*, or you might be classified as a "reactionary kook." Better not print too many items about the failure of cloud-seeding. Are you implying we should look to *other*, mayhap disreputable, means of rainmaking?)

To continue—May 1955 headlines: "Atomic Residues Pose a Challenge" (Now, that's a superb headline—"Atomic Residues Pose a Challenge"! Best thing to do is hunt around for another probe, or maybe another committee.) . . . "Action Urged in Smog Problem" (You mean to say somebody was actually urging *action* back then in the long-past *1955*? Another kook, no doubt!) . . . "Eight-Inch Snow in Britain Swirls in Mid-May Gale" (Chalk it up as another anomaly, like the eight-legged frogs—happens all the time. Eight-inch snowfalls are quite common in Britain in mid-May!)

Now, in that same May of 1955 we get this reassuring headline from the N.Y. *Times*: "AEC Sees No Peril in Nevada Fall-Out" (Well, let's be thankful for little favors—at least the *AEC* sees no peril. That's very gratifying indeed.) But wait—hang onto your hats! Look at this headline from the Associated Press

—same mid-May, eight-legged, eight-inch British snowfall month: "A-Test Burns Claimed in 3 Suits Against U.S." (Somebody is just trying to be a spoilsport about this atomic radiation pollution; or when the AEC saw "no peril," maybe they weren't looking at the right places or *faces*: "A-Tests Blamed for Face Cancer," comes a terribly disturbing headline from the AP); then another headline: "Geneticist Tells of Fall-Out Harm" (Cross him off as another kook!) But all's well that ends well in May 1955, with the boldly bracing headline from the redoubtable AEC: "AEC Minimizes Nevada Fall-Out" (Thank God somebody's there to *minimize* it. You really don't expect the AEC to *maximize* it, do you?)

Now, lest I be accused of bias (which I freely admit to), I shall keep my impertinent comments to myself and merely list the headlined items that I culled from the news media. We move to year 1957:

"Canadian Tests on Milk Show Fall-Out Harmless Since A-Tests" . . . "Radioactive Rainfall Ruins Japan Fruit" (Reuters) . . . "UN Body Warns on Radioactivity" . . . "Nations Warned of Perils Ahead From Rise in Atomic Pollution" . . . "Reds Set Off New A-Bomb" . . . "U.S. Troops to Eat Irradiated Food: Army Plans to Demonstrate Practicality of Atomic Preservative Process" . . . "Connecticut Votes Study of Radiation" . . . "Pauling Lists 10,000 as Fall-Out Victims" . . . "U.S. Opens Study of Radioactivity—extensive inquiry to include milk and water" . . . "Khrushchev reported slightly ill" . . . "12-Inch Snow Cripples New York City" . . . "Russia Launches New Series of Nuclear Tests" . . . "Strontium-90 Shows Sharp Rise in New York City" . . . "Radiation Fells 200, Under Treatment in Moscow Hospitals" . . . "Nevada Woman Exposed to A-Bomb Fall-Out" . . . "AEC to Sift Delay on Accident Report" . . . "Radioactivity in Milk Held No Peril, Says U.S. Public Health Service."

The following items appeared in the news media from June through October 1957:

"Four Geneticists Say Fall-Out Perils Future Generations at Congressional Hearings" . . . "Experts Divided on Fall-Out Issue" . . . "Government Scientists who tried to circulate a petition

against continued H-Bomb tests suddenly found their Constitutional right of free petition yanked from under them at National Institute of Health" (Drew Pearson) . . . "Atom Hysteria Called Reds' Goal" . . . "Soviet People Informed of Five Russian Tests" . . . "Air Pollution Move by U.S. Sought" . . . "U.S. Atomic-Test Area Calm on Fall-Out: Only a Few Persons Within 200-Mile Radius Show Concern" . . . "Scientists Stress Dangers of Fall-Out: No 'Clean' Bombs" . . . "H-Shot at Moon Urged by Dr. S. Fred Singer" . . . "The amount of [radium] on watches and clocks is so slight it has no ill effects" (*Life* magazine).

Now we move into year 1958:

"Rise in Strontium Noted in Humans: Nuclear Explosions Caused a 33% increase in the concentration of radioactive Strontium-90 in the human body" . . . "Atom Tests Called a Food Peril by E. Knopp, German agricultural scientist" . . . "Minneapolis Seeks Fund to Test Water: Radioactive readings as high as 180 micromicrocuries, made last summer in Minnesota drinking water" . . . "Dr. Libby Asks World Pact to Restrict Tests."

In March 1958 we find these headlines:

"Army Plan Lags in Food Radiation" . . . "Tokyo Air More Radioactive" . . . "AEC Accused in Fall-Out Data" . . . "Atom Ships Held Threat to Ports" . . . "Fall-Out Reports Stir Coast Area—tests find no evidence Soviet blasts contaminated San Francisco's water" . . . "Nuclear Tests Held Not a Health Peril" . . . "Fall-Out Secrecy Hit" . . . "Youths Most Susceptible to Nuclear Fallout—One-year-old children have highest bone concentration of radioactive Strontium-90, reported by Dr. J. Laurence Kulp of Columbia University." (A previous study by Kulp and others showed that children four years old and under had *ten times more strontium-90 than adults*—N.Y. *Herald-Tribune*.)

Then in March and April 1958 we read:

"Rise in Radioactivity Reported in Washington: Highest Levels of the Year" . . . "Red Tests Hike Denver Fallout: Radiation Jumped to Ten Times Normal" . . . "Hot Mud Scrubbed Off Rome Cars: Mud showed an appreciable degree of radioactivity" . . . "Latest Soviet A-Tests Believed 'Dirtiest' Ever" . . . "Dr. Libby Reports High Radioactivity Over New York a 'quirk of nature' " . . . "Northwest U.S. Has Highest Radioactivity in Nation" . . . "Japan Records High Radioactivity" . . . "Two Physicists Decry Fear of Radiation—'Much more danger from radium dial watches in the U.S. than from all the fall-out radiation of bomb tests.' "

Thus, more than ten years ago, I was clipping and noting worldwide items concerning radiation pollution. My work on the *Journal American Water Works Association* afforded additional opportunities to file away further material in this field, data that I did not use until I moved to Idaho, where the threat to life was too menacing to be ignored.

3 Practical Politics

Practical politics consists in ignoring facts.
Henry Brooks Adams

When my wife and I moved to Idaho Falls in March 1967, fighting a public battle with the Atomic Energy Commission was the furthest thing from my mind. After working for about a year and a half as military and city editor of the *Post-Register,* as I stated previously, I was appointed to head up a newly established agency designed to look after the needs of senior citizens of Bonneville County. I might mention here that my agency began a number of pilot projects which gained national recognition. Under my direction, the Eastern Idaho Special Services Agency instituted programs to aid the elderly and indigent, and organized volunteers to help county senior citizens; and after about eight months of operation, the agency became a participating member of the Bonneville County Community Chest, alongside such reputable agencies as the Salvation Army, the Boy Scouts, and the American Red Cross.

On October 12, 1969, the *Post-Register* carried the following page-one headline: "Governor Inquires About Wastes Disposal." Datelined Boise (AP), the story ran:

> Gov. Don W. Samuelson asked the Atomic Energy Commission Friday if it is making a continual review of the procedures used in disposing of radioactive wastes to be certain that there is no danger of contamination.
>
> Samuelson directed his letter to Dr. Glenn T. Seaborg, AEC chairman.
>
> The governor said he acted because of a letter received from the operator of a trout farm which uses water from the Snake River, who questioned whether there might be contamination from wastes buried in Eastern Idaho.
>
> *"Even though I am advised there has never been any evidence of contamination of water over the 15 years that the burial of the waste has occurred,"* Samuelson wrote, "is the U.S.A.E.C. continually reviewing the matter and exploring ways and means for the ultimate in safety standards? [Italics added]

"Although currently used practices have been entirely successful thus far in preventing contamination, should some unforeseen or unexpected leakage occur, the resulting disaster would be enormous and irreparable.

"Because of the continuing concern in the public's mind, I would appreciate an early response on the current attention which the U.S.A.E.C. is devoting to this matter."

This article posed some interesting questions. The operator of the trout farm mentioned here is Robert A. Erkins, president of the Snake River Trout Co. of Buhl, Idaho, the largest commercial trout farm in the United States, if not in the world. Buhl is roughly 185 miles southwest of Idaho Falls (or more than 150 miles from the National Reactor Testing Station and the nuclear-waste dumping grounds near Arco in the desert).

Why was Mr. Erkins concerned about the safety of his baby trout nearly 150 miles from that desert "hot-waste" dumping grounds? For some 15 years the AEC had been pouring millions of gallons of radioactive waste material into that desert area which is located *immediately over* the Snake River aquifer. The Snake River aquifer is an underground body of fresh water of incalculable size, which in turn feeds most of the Northwest. No one can say with certainty exactly where that water flows; subsidiary underground streams may run in many directions, affording a source of water supply to countless public and private wells and watersheds. As we shall see later, people have been killed in Italy as a result of drinking radioactive water from a contaminated well. Let us also remember those eight-legged frogs.

It was reassuring to read in this article that the Governor of Idaho had been advised "there has never been any evidence of contamination of water over the 15 years that the burial of the waste has occurred."

As an ex-newsman, however, I was well aware that oftentimes, to allay public fears, politicians (including scientific politicians) will give out a news release pooh-poohing the possibility "of the slightest danger whatsoever" from some "research" which they have an interest in. For years the DDT pushers had been ridiculing the potential hazards of that chemical killer. In this category also belongs the quiet, unmourned demise of the much-touted sulfa drugs. It wasn't too long ago that the Chemical Warfare boys denied any connection between their "research" at the Dugway Proving Grounds in Utah and the death of thousands of sheep; and despite continued denials of liability (in the form of news releases), the Army did nevertheless pay the sheep-herders for their dead animals and polluted grazing lands.

Now Mr. Erkins was inquiring about the safety of the waste-disposal methods being used in southeastern Idaho. Some of that radioactive waste material will remain "alive"—that is, insidiously active—for *thousands of years.* The steel containers holding these boiling poisons have been known to spring leaks. *Eventually they must rust out!* It is scientific poppycock for anyone to state "with certainty" that the burial of such wastes "poses no danger"! Who knows what is actually going on below the surface of the ground? Particularly unnerving is the fact that this radioactive graveyard is located within an earthquake-prone region.

Hindsight is very easy. It eases the conscience. On May 20, 1965, for example, the New York *Times* gave the public reassurance from the U.S. Food and Drug Administration that artificial sweeteners—including cyclamates and saccharin—were "safe" at "present-use levels."* Only a few years later, cyclamates were suddenly banned! Meanwhile, of course, there must have been damage done to thousands, if not millions, of people. Again, on October 28, 1965, the New York *Times* noted that schoolchildren in a Utah county might have an abnormally high incidence of small lumps on their thyroid glands. The U.S. Public Health Service, however, reassured us that "an investigation was under way." May we dare to suggest even the slightest connection between the nuclear tests conducted in southeastern Nevada and glandular lumps in Utah children?

---

*March 10, 1971—The U.S. Dept. of Health, Education & Welfare announces that amphetamines (pep pills) in "properly administered doses" aid schoolchildren with "behavioral problems." I suggest this slogan for teachers: "Have problem children? Don't hug 'em—drug 'em!"

So Mr. Robert Erkins, 150 miles away from a lethal dumping ground, is quite rationally concerned over the safety of millions of his baby trout. Those trout represent Mr. Erkins' ability to survive economically. Having spoken and corresponded with Bob Erkins, I am, moreover, convinced that he is also a humanitarian whose concern encompasses more than economic survival. He wants to be assured that there is no threat to *Life*—his life and the lives of his family, friends and neighbors, as well as to the community at large. And he has a right to this assurance.

Yet no scientist worthy of that title could give absolute assurance of such safety. If he did so, he would be an irresponsible fool. No one can predict an earthquake, or even a slight temblor, that could dump those millions of gallons of radioactivity into major water courses. No one can predict torrential rains and floods (and southeastern Idaho has had them) which could disrupt the "peaceful containment" of those radioactive contaminants and send them slurrying into rivers, streams, and God knows whose well and tap water!

What prevents most people from speaking out *beforehand* about the evident or possible threat to Life posed by new processes or inventions? This simple question has puzzled me for years. The only answer I could uncover is *fear*! Fear of what? Fear of being kicked off their neat, quiet ways of living. Fear of the possible loss of income. Fear of rocking the boat. Fear of making waves. Fear of losing their prestige. Fear of everything except the deepest, most rational fear: *fear for the safety of their very lives*! In short, I can only conclude that civilized man, as a *distortion* of rational Life, has lost or strangled (or drugged to death) his basic instinct for survival. Certain segments of society, for example, blatantly proclaim, "Better Red than dead." Others just as blatantly (albeit silently) proclaim, "Better radioactive than broke!"

In the early 1950s I was working as a sales engineer for a manufacturer of medical equipment. One summer, our company displayed its latest equipment at a convention booth in the Cow Palace of San Francisco. Next to our booth, which I manned during the day, was a booth occupied by a manufacturer of electronic instruments. In the course of the week spent there I became friendly with the electronics engineer who occupied that adjacent booth. I asked him about the possibility of TV sets giving off X-rays. He said that *all* TV sets give off "soft," or low-voltage, X-rays, but these were of "minor significance." Color TV sets, he acknowledged, gave off much more harmful doses—in both quantity and quality. I asked the engineer why these facts were not generally known by the public. He looked at me with some indignation. "We're in business to *sell merchandise,*" was his answer.

Five years later I was teaching school in New York State, when I read a news article connecting the close viewing of television by children with an alarming increase in glaucoma. I called this article to the attention of the school's principal, suggesting that all children in our district be notified of the possible dangers of radiation harm from TV sets. The principal refused to do anything about the matter. But I could not ignore my responsibility to some 250 students in my care, so I took it upon myself to warn them of this potential threat to their lives. As a result of my action, I was officially reprimanded for exercising "poor judgment" in discussing "extraneous" material with my students and for alarming them unnecessarily.

My concern for the safety of my students led to my dismissal in 1959. Evidently the authorities considered me a greater threat than nuclear radiation.

Ten years later the following headline appeared in the Idaho Falls *Post-Register*: "Limits Due on TV Rays." The news story, bylined by Frank Murray out of Washington, D.C., stated in part: "The new rules, first to seek limits [by the Federal Government] on X-radiation from television sets, are scheduled to be published . . . in the Federal Register. . . . They stem from a law enacted last year following disclosures of erratic and sometimes high emission of X-rays from color TV sets. Government scientists have told Congress there is no evidence of direct harm to humans from radiation emitted by color television sets. *But they also said that radiation accumulates in the body so any exposure brings a person closer to the unknown point where genes are mutated, possibly causing damage to future generations.*" (The article appeared Oct. 22, 1969—italics added.)

The blindness that may result to only *one* child is cause for alarm. But what of the scientific blindness that can at one and the same time state "there is no evidence of direct harm from radiation emitted by color TV sets" while maintaining "*any* exposure brings a person closer TO THE UNKNOWN POINT where genes are mutated!" If you don't know at what point genes are mutated by radiation, is it sanity or insanity to continuously *increase* the natural radiation now in our atmosphere? Is there a possibility that hundreds, if not thousands, of newborn babies each year are suffering retardation or malformation due to the relentless increase of biospheric radiation?

The gigantic schizophrenia of the nuclear-power industry is glaringly evident in this double-standard, double-visioned, double-thinking attitude. On the one hand, we are told that emissions from nuclear smokestacks, from nuclear waste water, from the radioactive burial grounds, from radium-dial watches, from high-frequency ovens, from "scatterings" of thousands of radar installations, from the microwave stations—all these sources are merely emitting minute, "barely detectable" quantities of radiation which can in no way be construed as harmful to humanity. On the other hand, as we have seen, these same sources of myopic information tell us that *they don't know when man's genes are mutated*! We have only been confronted thus far with eight-legged frogs. What will it take, eight-legged *human babies*? Weren't thousands of malformed thalidomide monstrosities *enough* to give the atomic fraternity pause for deep, soul-searching concern? Apparently not.

From the public relations offices of the atomic energy interests the pap of safety and reassurance continues to pour forth, like the fabled salt mill below the surface of the ocean. Anyone who challenges the atomic industry's right to pollute the planet and endanger its inhabitants is immediately classified a kook, crank, or incompetent nut—even when such opposition is raised by highly qualified nuclear scientists themselves—those few who have had the guts and humanity to cry wolf when they see danger. But in the main, those with vested monetary interests in the atomic industry continue to yell, "It's quite safe! There is no real risk to life!" Maybe. But if that is so, how come *every* standard automobile policy, and every homeowner's insurance policy, *absolutely excludes* insurance coverage against radiation loss to humans and property? Read the fine print in your policies. Mr. and Mrs. American car owner and homeowner are insured against nearly every form of peril, from flying objects falling from the sky to vandalism by Mrs. Murphy's brats. Yet the insurance companies exclude coverage from so relatively "safe" a hazard as nuclear radiation! I am not talking about damage from nuclear war—such war damage is another, separate exclusion. Your policy excludes damage *from any and every source* of nuclear radiation. Insurance companies are in business to make money by insuring against the probabilities of certain perils injuring or destroying life and property. These policies are tacitly stating that the risks from nuclear damage *are far greater* than the risks from, say, your boiler exploding or your house catching fire. Don't take my word for it: read your policy!

4 The Nightmare

Radioactivity represents one of the worst, maybe the worst of all poisons. And it is manufactured in astronomical quantities as an inevitable by-product of nuclear electricity generation.

From *Poisoned Power*
by John W. Gofman, M.D.
and Arthur R. Tamplin, Ph.D.

A short time after Governor Samuelson's letter of inquiry to the AEC was printed in the *Post-Register*, a local news commentator referred to the Governor's concern in a radio broadcast. The commentator characterized the Governor's concern as being that of an elderly grandmother who seeks reassurance that she is not really ill.

All public figures are fair game for the news media, which seldom miss a chance to take potshots at anyone who happens to stick his neck out. To me, however, the commentator's sarcasm was out of place. Plutarch remarked: "They that were serious in ridiculous matters would be ridiculous in serious matters." The matter at hand was far too serious to be tossed off and ridiculed by anyone.

I gathered together the notes and memoranda I had been collecting for many years, and wrote the following pamphlet, which I sent to Idaho's congressional delegation, Governor Samuelson, members of the Idaho state legislature, the Bonneville County commissioners, prominent and influential citizens of Idaho Falls, and the local news media. . . .

NUCLEAR NIGHTMARE

Recent and continued testing of nuclear devices in the West, and the dumping of nuclear waste materials in Idaho, particularly at the National Reactor Testing Station installation 50 miles west of Idaho Falls, are focusing public attention on the possible hazardous effects of such actions.

From 1962 to 1966, I served on the editorial staff of the *Journal American Water Works Association*—the last two years as managing editor of this most respected scientific publication in the field of public water supply. I was therefore in a unique position to read about and to follow developments in our national and planetary environment.

I felt obliged, therefore, to call to the attention of public officials some facts and conclusions regarding nuclear materials which are not generally known.

Regardless of personal affiliations, it appears that many responsible scientists and officials in high places have voiced—and are continuing to voice—grave misgivings with regard to the testing and use of nuclear devices and the disposal of nuclear wastes. The first chairman of the Atomic Energy Commission, David E. Lilienthal, described a proposal to construct a nuclear power plant in Queens, N.Y., as a "very risky business."* Mr. Lilienthal told the Senate-House Atomic Energy Committee in Washington, D.C., that he "would not dream of living in Queens" if such a plant were built there. As reported in the New York *Times*, April 5, 1963, the former AEC chief called the Queens proposal "a very risky business" for the residents of the Long Island area because of the problems of atomic-waste disposal and possibilities of accident or sabotage.

Such a statement issued by the then chairman of the Atomic Energy Commission is cause for some concern. The U.S. Atomic Energy Commission in its publication *Major Activities in the Atomic Energy Programs*[1] provides us with the names and locations of *hundreds* of reactors—operable, in construction, or planned. It is quite clear that escape from this nuclear threat is all but impossible.

Writing in the Stanford University periodical *Stanford Today* (autumn 1962), the world-renowned Professor Rolf Eliassen of Stanford's Department of Civil Engineering had this to say:

---

*At the time this was written, I was unaware of the insurance-policy exclusions which give no protection against nuclear radiation to life and property, thus demonstrating how really "risky" a business the nuclear industry is!

> To these factors of potency and hazard to health we must add the problems of tremendous quantity and the long life of radioactivity. Millions of gallons of these waste waters will be produced each year at the chemical plants which will be built to process spent fuel elements. These plants are located in remote areas, and storage of the liquid wastes is accomplished at the present time in large underground tanks. Over 70 million gallons are already stored at Hanford, Oak Ridge, Idaho, and Savannah River plants of the U.S. Atomic Energy Commission. Great quantities must also be stored in France, Russia, and Great Britain. . . . Even after a year of decay and virtual disappearance of the short-lived isotopes, the radioactivity which will remain in these waste waters will amount to about 100 billion curies a year at the predicted rate of power production in 1990. . . . *All of the oceans of the earth do not contain enough water to dilute these billions of curies of wastes to man's level of tolerance.* [Italics added]

One hesitates to conjecture on the possible effects an earthquake might have on such underground sources of "contained" radioactivity. On September 4, 1965, columnist Jean Pearson of the Detroit *Free Press* reported the death of 16 villagers near Rome, Italy, from cancer induced by drinking water from a newly dug well. Investigators found that the water contained 4,400 microcuries per liter. The "safe" dose is 100 microcuries per liter.

Which brings up an interesting point with respect to determining what exactly *is* a "safe" dose of ionizing radiation. In its 1962 Drinking Water Standards,[2] the U.S. Public Health Service states: "The effects of human radiation exposure are viewed as harmful, and any unnecessary exposure to ionizing radiation should be avoided."

According to the U.S. Atomic Energy Commission: "No single source of data directly yields the relationship between the physical dose of ionizing radiation and the clinical effect in man. Hence, there is no complete agreement concerning the effect associated with a specific dose or dose range."[3]

Thus, although there is no complete agreement regarding the effects of any specific dose of radiation on the human organism, the scientists do at least acknowledge *the hazard to life from all forms of ionizing radiation.*

James Miller, project officer of the USPHS Dental Radiological Health Project, stated that "the conservative assumption is that the somatic and genetic effects resulting from radiation insult [!] have no threshold and that a direct linear relationship exists between dose and biological effect."[4]

Another comment on the harmful effects of ionizing radiation worthy of note comes from Dr. E. Windle Taylor, Director of Water Examination of the Metropolitan Water Board of London, England. Dr. Taylor wrote:

> The harm from radiation is two-fold, namely, a localised or somatic effect, upsetting the actual cells of the body and leading to such conditions as leukemia, anemia, bone tumors, dermatitis, and cataract, and a long-term or genetic effect resulting from the action of ionizing radiations on the chromosomes in the reproductive cells. This gives rise to such effects as dwarfism, hemophilia, metabolic and mental diseases in succeeding generations.
>
> By the increasing use of artificial radiation we are confronted with a new and very insidious effect on health.[5]

Two facts emerge sharply and unmistakably from statements such as the ones quoted and from many similar statements scattered throughout the literature. First, ionizing radiation from whatever source and in any amount is dangerous to life. And second, there already exists on the earth and in the atmosphere sufficient radioactivity to cause the gravest concern on the part of responsible scientists everywhere.[6]

Despite these sobering facts, scarcely a week goes by without the detonation of a nuclear device. Conservative estimates of the combined number of planetary explosions during the past ten years indicate *thousands* of such detonations have occurred—with no end in sight!

Despite the AEC's repeated assurances of "safety," occasionally one reads information to the contrary. On May 2, 1963, for example, the New York *Journal-American* reported that for the *seventh consecutive month,* the town of Palmer, Alaska, received a fallout of iodine-131 that was above the "acceptable health risk" level!

Such phrases as "acceptable health risk level" make us ponder over nuclear double talk. We know there is *no* safe dose threshold of ionizing radiation. Who, then, decides when the "health risk" is "acceptable" or "unacceptable"? Not the people of Palmer, who risk the loss of their dairy business, if not their lives. Who decides whether freshwater clams taken from the Tennessee River are "acceptable" or "unacceptable"? It is well known that clams concentrate strontium-90 in their shells.[7] It is equally well known that strontium-90 has an affinity for milk and that ionizing radiation is being detected in almost every food now ingested by man and beast.[8]

And so one begins to wander deeper into this nuclear nightmare, searching for some glimmer of hope, an island of safety, a straw of sanity. We recall the devastation caused by the nuclear explosions at Hiroshima and Nagasaki, and we turn to the radiologists, the *experts* in radiology, for a word of disapproval of this dangerous *abuse of power.* The radiologists cannot help us. They too appear to be trapped in this nuclear nightmare. We learn that THE DEATH RATE FROM RADIOLOGICALLY INDUCED LEUKEMIA IS FOUR TO FIVE TIMES GREATER IN RADIOLOGISTS than in non-radiological physicians![9]

Even the iron curtain cannot bar the onslaught of ionizing radiation, and we learn of "considerable functional disorders" following "therapeutic" radiation treatments in a study covering a 10-year period.[10] The Japanese, too, have not been immunized against this deadly power: a study details 15 new cases of radiation cancer of the neck following irradiation of non-cancerous diseases![11]

Because of the tremendous increase in the industrial use of ionizing radiation and the concomitantly higher risk to the health of workers who are thus exposed, studies were being made back in 1962 *to include leukemia as a compensable radiation injury among such workers.*[12] The higher incidence of sterility, stillbirths, and infant deaths among X-ray workers has also been known for years.[13] And the use of fluoroscopy during surgery has made every surgeon aware of the danger of radiopathies (diseases induced by radiation)[14] both to himself and to his patients. And yet ionizing radiation is one of the most widely used "treatments" for cancer.

In the chemical and drug laboratories of the world, compound after compound is being tested in the vain hope of finding some counteragent to the life-damaging effects of ionizing radiation which these same laboratories hail as a boon to mankind. One study, for example, describes 1,200 compounds tested against radiation injury.[15] The toxicity of each compound is described.

And while ionizing radiation increases daily in the atmosphere, in the ground, under the ground, and in our rivers and oceans, scientists continue methodically to study its effects. They feed irradiated beef to dogs;[16] and, of course, if dogs can eat it, why not man?[17] However, here and there, courageous and concerned scientists manage to raise a voice of protest through this nuclear smog. "Radiation Called Lethal," reported the New York *Times,* December 9, 1965:

> Three Cornell University scientists cautioned that food preserved by nuclear radiation might be lethal. . . . The Defense Department and the food industry have both been testing radiation as a preservative. . . . The warning was issued by Professor Frederick C. Steward and two researchers, Dr. Richard D. Holsten and Dr. Michiyasu Sugii.

Summary and Conclusions

In view of the grave statements made by highly qualified people in the field of nuclear physics, it is apparent that the atomic energy interests have embarked upon a highly dangerous course of research and development. The recent sheep disaster attributed to chemical warfare research in Dugway, Utah, indicates that those directly involved in experimental research are not always aware of immediate or long-range consequences. And despite repeated denials of responsibility to the sheep owners, reparation was nevertheless made by the Federal Government to those sheep owners.

It is necessary also to recall the thousands of monstrosities whose lives were permanently maimed by thalidomide!

What kind of reparation would compensate a wife whose husband has contracted leukemia? Where are the comparative studies indicating the incidence of leukemia or cancer in workers in nuclear plants? Where are the published reports demonstrating the safety of drinking water in towns situated near nuclear reactors? Where are the published reports indicating that the incidence of birth defects in families living in close proximity to nuclear reactors is lower than, equal to, or higher than the incidence of birth defects in families not so located?

The development of the atomic energy industry, like the development of any new industry, involves mankind in certain risks. Many thousands of coal miners were killed in "accidents" associated with their work before widespread, responsible safety measures were instituted. The history of industrial accidents indicates that, unfortunately, those who have a vested interest in such industries are too often reluctant or unwilling to take the time and spend the money for adequate human safeguards, until forced to do so by Federal, state, or local government.

The crucial question is this: *Is the nuclear energy industry sufficiently aware and concerned about the welfare of human life*? Many of the sources quoted in this paper strongly indicate that *this is not the case*! It then becomes the responsibility of the citizens themselves to insist that their local, state, and Federal governments take immediate and forthright steps to insure the welfare and safety of the people. This can be accomplished only by establishing suitable watchdog committees composed of *impartial* citizens who have NO VESTED INTEREST, either directly or indirectly, in the atomic energy industry. Members of such a watchdog committee must be of unimpeachable character and integrity, responsible solely to *the people* of the state they serve, under the direct administration of the governor.

Nothing should be more sacred to human society than human life. Contempt for life in the desperate pursuit of financial gain is a shabby and despicable sellout of moral principle. It is my hope that responsible representatives of the people will take prompt and courageous action to insure the welfare and safety of our state and society, as to both present and future generations.

Notes

1. US Atomic Energy Commission, *Major Activities in the Atomic Energy Programs—January-December, 1962* (US Govt. Printing Office, Wash., D.C.).

2. USPHS, *Drinking Water Standards, 1962* (US Govt. Printing Office, Wash., D.C.).

3. S. Glasstone, ed., *The Effects of Nuclear Weapons* (US Atomic Energy Commission, US Govt. Printing Office, Wash., D.C., Apr. 1962).

4. J. W. Miller, "Summary of State Dental Radiological Health Activities," *Radiological Health Data,* vol. 4, no. 1 (Jan. 1963).

5. W. E. Taylor, "Radioactivity in Waters and Treatment of Radioactive Waters," *Proceedings.* vol. 2, 5th Congress, International Water Supply Assn., Berlin (1961).

6. *Radiological Health Data,* vol. 4, no. 3 (Mar. 1963) (US Dept. of Health, Education & Welfare, Wash., D.C.).

7. D. J. Nelson, "Clams as Indicators of Strontium-90," *Science,* vol. 137 (July 6, 1962).

8. *Radiological Health Data,* vol. 4, no. 2 (Feb. 1963) (US Dept. of Health, Education & Welfare, Wash., D.C.).

9. H. C. March, "Leukemia in Radiologists, Ten Years Later, With a Review of the Pertinent Evidence for Radiation Leukemia," *American Journal of Medical Science,* 242:137 (Aug. 1961).

10. M. P. Domshlak, et. al., "Late Observations on Persons Subjected to the Effect of Radiation," *Nuclear Science Abstracts,* no. 11477 (May 31, 1962).

11. H. Yamashita et. al., "Radiation Cancer: Cancer Caused by Irradiation of Non-Malignant Diseases of the Neck and Pharynx," *Nuclear Science Abstracts,* no. 25201 (Oct. 1962).

12. R. R. Newell, "Leukemia as Compensable Radiation Injury," *Archives of Environmental Health,* vol. 5, no. 2 (Aug. 1962).

13. T. Kitabatake, "Sterility, Stillbirth, Infant Death, and Sex Ratios of Off-Springs of X-Ray Workers," *Nagoya Journal of Medical Science* (Nagoya Univ. School of Medicine, Japan), 23:277 (1960).

14. Hector Marino, "Radiopathies in the Hand of the Surgeon," *Nuclear Science Abstracts,* no. 32818 (Dec. 1962).

15. John Doull, et. al., *A Survey of Compounds for Radiation Protection* (School of Aerospace Medicine, Brooks AFB, Tex., 1961).

16. Elwood F. Reber, et. al., "The Effects of Feeding Irradiated Beef to Dogs," *American Journal of Veterinary Research,* 23:74 (Jan. 1962).

17. G. D. Gernon Jr., and R. W. Seaton, "Irradiated Meats: (I) Storage Stability of Cooked and Raw Meat," *Journal American Dietetic Assn.,* 41:20 (July 1962).

5 *Shock Waves*

All I know is just what I read in the papers.
Will Rogers

My documented pamphlet "Nuclear Nightmare" was mailed out in the second week of September 1969. It was read in its entirety over a local radio station, followed by a request for an interview which I granted to a local television station. The Idaho Falls *Post-Register,* however, did not consider the pamphlet worthy of printing, and it was never alluded to by my former boss, executive editor Robb Brady, who took a personal interest in atomic matters. Nevertheless, despite the outward calm and silence, I knew that my pamphlet was being discussed behind closed doors—by private citizens as well as by interested members of the AEC.

On October 8, 1969, an article appeared in the *Post-Register* which was of great interest to me; and so I wrote the following open letter to Governor Samuelson.

Dear Gov. Samuelson:

I see by today's Post-Register that James Agee, regional director for the Federal Water Pollution Control Administration, admitted to you that "Radioactivity in waste waters which reach the aquifer underlying the Snake River Plains *apparently is so slight as to pose no danger.*"

So, after many years of denying ANY radioactive waste is polluting our underground water supply, we finally get an admission of "slight" pollution.

Gov. Samuelson, please do not be lulled into a false sense of security by such "authoritative" disclaimers of danger. In discussing radioactive pollution of the Columbia River, Dr. Albert Schweitzer stated: "The radioactivity of the river water was insignificant. But the radioactivity of the river plankton was 2,000 times higher, that of the ducks eating the plankton 40,000 times higher, that of the fish 150,000 times higher . . . and the egg yolks of water birds more than 1 million times higher." [As quoted in *Atomic Suicide;* see Bibliography.]

Thus, Dr. Schweitzer makes it crystal clear that the entire life-cycle, from water to plants, to fish, to fowl, and finally, man, are all involved, with the radioactivity being *concentrated* and increased in each successive link of that life-cycle chain!

Radiation poisoning hits each person IN HIS WEAKEST SPOT! Dormant illness are "triggered," and since the general symptoms are so similar to "virus X" (chronic weakness, oxygen and water hunger, diarrhea leading to enteritis, etc.) many people believe they "just have a virus."

Please, Gov. Samuelson, do not abandon your stand on insuring the safety of Idaho's water supply. A "little" radioactivity is just exactly like a "little" cancer—it can and will kill us.

Respectfully yours,
JEROME EDEN

The letter was published in the *Post-Register* on October 15, 1969. On that same date, the *P-R* carried these items of note: "Teller Says Reactors 'Potentially Dangerous' "—a Seattle (AP) story:

> Dr. Edward Teller, one of the chief architects of the H Bomb, says "nuclear reactors are potentially dangerous."
>
> Teller said Wednesday he would like to see reactors in a place where sabotage would not be probable, such as under the ocean or deep under ground and cooled by ocean water. Referring to power from nuclear reactors, Teller said: "If you want more electricity, you'll do it by heating up rivers. There is danger for some fish."

On the same day, an Associated Press story out of the United Nations, N. Y., stated: "Atmospheric nuclear tests since 1966 contributed substantially to the amount of long-lived radioactive nuclides deposited on the earth's surface." No mention was made here that such tests also "contribute substantially" to the amount of long-lived radioactive substances which filter down upon human beings, deposit themselves upon hair, hands, clothing, are ingested on foodstuffs so contaminated, and are carried silently into our very tap water.

Again, on October 15, 1969, the *P-R* carried this headline: "AEC Careful About Waste." Datelined Boise (AP), the article stated:

> A nuclear scientist said . . . the Atomic Energy Commission

> is "very careful" about wastes, but added it is good the issue is being aired.
>
> Dr. Willard Frank Libby, Nobel Prize winner who recently was named the state's science advisor, made the remarks during a news conference with Gov. Don Samuelson.
>
> When asked about AEC waste disposal policies, Libby said, "The general rule is that the AEC is very careful. But if there is a question about it, I think it should be brought out in the open. . . ."
>
> Libby, a former member of the Atomic Energy Commission and now a professor of chemistry at UCLA, said Idaho is a "wonderful, unspoiled state. Keep it that way."

All I can say in response to Dr. Libby is, "We intend to, sir. We intend to!"

On October 16 the following headline appeared in the *P-R*:

ARMY'S SHEEP-GASSING
COST COOL $1 MILLION

The article detailed the death of 6,400 sheep "after a spray tank malfunctioned during an aerial test of the nerve agent VX, the Army said. Extensive investigations finally revealed that the sheep probably ingested very low dosages of the agent VX." (Note: The 6,400 dead sheep should be happy to learn they ingested only "very low dosages of the agent VX"!)

"Another $198,309 was paid for claims regarding temporary loss of land use," the article stated; but the last paragraph of the story added this disclaimer: "Witnesses denied reports that the accident caused much of the affected land to be retired permanently from use."

Incidentally, if you're wondering who is footing the bill for this death-dealing goof, look in the mirror, Mr. and Mrs. American Taxpayer!

On October 22, 1969, the *P-R* blazed out a five-column headline: "AEC Tests Show No Ray Problem in Area." There again was Mr. Robb Brady at the bridge, confidently reassuring his readers that nothing was or could go amiss. In the second column of this lengthy article of reassurance the following sentences appeared: "On the other hand, the average yearly concentration of gross alpha radioactivity *in Idaho Falls is 34 per cent higher than that at the site.* Investigations are underway to find the source of this higher gross alpha concentration in Idaho Falls." (Italics added.)

So the gross alpha radioactivity in Idaho Falls is 34 per cent higher than it is at the NRTS site; and they don't know why! It couldn't be because the prevailing winds are from west to east *and Idaho Falls is east of the site!* (Besides the burial of oceans of radioactive wastes, the NRTS is the *testing station* for many different types of nuclear reactors which spew their "harmless, low-level radioactivity" into the atmosphere!)

U.S. Senator Len B. Jordan of Idaho wrote me in response to my criticism of the waste-disposal practices of the AEC in Idaho. Senator Jordan invited me to testify before the Joint Committee on Atomic Energy which was preparing hearings in Washington, D.C. I sent my testimony ("Nuclear Nightmare") to the congressional committee. Receipt of the material was acknowledged, but it was never included in the three volumes of the hearings.

On November 7, 1969, this headline appeared in the *P-R:* "Reports Said Misrepresented . . . Health Physics Group Stresses Radiation Safety Here Assured." This article, disseminated by the Eastern Idaho chapter of the Health Physics Society, raked me over the coals for my criticism of waste-disposal practices and for my stand on atomic pollution as given in "Nuclear Nightmare."

The article accused me of "choosing to ignore published information" demonstrating the safety of AEC practices. "We would like to comment on a few of Mr. Eden's specific remarks," the article stated. "Although he [meaning me] has generally quoted correctly, he has distorted the meaning or implication of the quotes, and, consequently, has magnified the actual hazard of radiation."

Thus I am publicly accused of "distorting the meaning or implication of the quotes," which I cited in "Nuclear Nightmare." All right—it is one thing to make accusations; it is another thing to prove them! What are the facts?

The article states: "Mr. Eden contends that radiologists die from leukemia 'at a rate 4 to 5 times greater than non-radiological physicians.' This may have been true of radiologists practicing prior to the 1920's before the effects of large doses of radiation were recognized."

Now, the Health Physics Society, which released this attack upon me, had a copy of my pamphlet "Nuclear Nightmare." First of all, "Mr. Eden" is not contending anything! Mr. Eden is merely quoting this conclusion from reference no. 9 given in the pamphlet by H. C. March, who wrote a paper published in the *American Journal of Medical Science,* August 1961. The paper is titled "Leukemia in Radiologists, Ten Years Later, With a Review of the Pertinent Evidence for Radiation Leukemia." As we can see, the paper was not published in the 1920s, as the Health Physics Society would have us believe, but in August 1961! Now, *who* is distorting the meaning or implications?

The next criticism of my veracity is written thusly:

"Mr. Eden relates several injuries that *he attributes* to radiation. The article which Mr. Eden says reports '15 new cases of radiation cancer of the neck' also states that 'it was not confirmed that previous irradiations were the carcinogenic cause.' Mr. Eden's statement that these were radiation cancers requires a determination that the authors of the article were not willing to make." (Italics added.)

Again, it was not *I* who related "several injuries attributable to radiation of the neck." All I said was, "A study details 15 new cases of radiation cancer of the neck *following irradiation of noncancerous diseases!*" I never used the word "injuries." The article referred to is cited in my pamphlet as reference no. 11, written by H. Yamashita et. al., entitled "Radiation Cancer: Cancer Caused by Irradiation of Non-Malignant Diseases of the Neck and Pharynx." If Dr. Yamashita uses the words "Cancer Caused by Irradiation," etc., in the title, it seems to me that he has certainly made the determination that these were radiation cancers! Again, who is distorting meanings or implications?

At the risk of boring the reader, let us look at one final criticism the Eastern Idaho chapter of the Health Physics Society leveled at me:

"Mr. Eden has also stated that many area residents have complained to him of 'the same general symptoms, chronic fatigue, unusual thirst, wandering aches and pains, malaise and nausea.' These are all symptoms associated with radiation sickness caused by extremely large doses of radiation. These effects could not be caused by the doses area residents could have received."

From the distant heavens comes the bleating of 6,400 sheep who dropped dead after ingesting "very low dosages of the agent VX." So we are back again to the argument of "minimal, minute doses—far below acceptable health-risk levels." Atomic radiation, as we shall see later, has a *triggering influence.* It takes only a three-pound pull of one finger to explode a .44-magnum cartridge. This same "minimum" three-pound pull can send the .44-magnum projectile speeding through the air at the rate of *1,470 feet per second.* Thus a "triggering" three-pound pull activates enough mechanical energy to blow a hole through the block of an engine. The human organism is far more complex and delicate—all tissues and organs function in a carefully balanced equilibrium and *harmony.* Knock off just one set of tissues or one organ anywhere in the body and you destroy that delicate balance throughout the entire organism. I'm sure the health physicists must know those facts. If you want to disrupt an entire symphony orchestra, you don't have to kill any of the musicians. All that is necessary is to knock off key the E string of any string instrument in that orchestra—the result will be utter chaos!

Radiation "insult" (to use the terminology of the health physicists) does not occur in a one-to-one relationship. In Nature, one plus one does not necessarily add up to two; in fact, more often than not it can result in twenty-two, or twenty-two thousand. One male rabbit plus one female rabbit *may result in a million rabbits!* One lightning bolt (or careless camper) plus a thousand tinder-dry acres equals a devastating conflagration. *Nature is functional,* not mechanical.

On November 24, 1969, the *P-R* carried the following headline: "AEC, Health Board Split on Radioactive Issue." The article began: "The Idaho Department of Health and the Idaho Atomic Energy Commission appear to be heading even more on a direct collision course over the disposal of radioactive waste.

. . ." In this article the Idaho Board of Health recommended to the AEC that the following measures be instituted:

> (1) Begin immediately to initiate a program of planning and implementation to the end that all new facilities constructed at the National Reactor Testing Station will have suitable waste disposal capabilities for handling radioactive liquid wastes without resorting to disposal directly into the ground water. [So far as I know, this was the first public indication that radioactive waste water was being introduced *"directly into the ground water"!]*
>
> (2) Establish a reasonable schedule for the replacement of all existing underground disposal systems with appropriate above-ground treatment facilities acceptable to the Federal Water Pollution Control Administration and the Idaho Board of Health. [Does this mean that present above-ground treatment facilities, *assuming there are such,* are *"unacceptable"* to the Federal Water Pollution Control Administration and the Idaho Board of Health? If so, why isn't someone screaming bloody murder?]
>
> (3) Formulate plans *for the eventual elimination* of the long-term storage of solid radioactive wastes and highly toxic radionuclides above the Snake River aquifer.

The above recommendations were published in November 1969. It is now three years later. To the best of my knowledge, the storage of millions of gallons of radioactive wastes continues unabated. However, I did receive some encouraging news from Idaho Senator Frank Church, who has often spoken out about the possibility of hazards to Idaho citizens from AEC practices. Senator Church believes that the present waste-disposal practices above the Snake River aquifer will be discontinued and the wastes removed . . . in possibly ten years!

6 The Red Shadow

Free men are those who refuse to yield under sentence of death what they are ready to yield of their own free will.

Wilhelm Reich

As an American citizen and a writer, it gives me no personal satisfaction to indict contemporary scientists, in the United States and throughout the world, for pursuing research which can only lead to catastrophe. To continue with further testing of nuclear devices, to expand our nuclear programs with the addition of nuclear reactors dotting our nation like so many time bombs set to explode, to continue to pollute our earth, air, water—and every living thing contained therein—with increasingly deadly amounts of radiation is equivalent to national and global suicide.

Our scientists have created a Frankenstein that threatens to destroy its creators. More and more every day, one hears the voices of concerned and sincere nuclear physicists, engineers, biologists, geophysicists, and laymen who are asking, nay, demanding, that nuclear research be discontinued—*right now!*

Nevertheless, our dilemma is not a simple one. We need our nuclear know-how to maintain a position of security in a hostile world. A rational man does not hand over his weapons to a band of threatening marauders, or dump them into the ocean and thus render himself completely defenseless. When one lives in bear country, one carries a bear gun.

Today we are still very much threatened by bears and dragons—Red Bears and even Redder Dragons. Despite the pictures of "Grandpa Khrushchev" appearing repeatedly on the cover of *Life* magazine, Red fascism still poses *the greatest single organized threat to the welfare of democratic peoples everywhere!*

What, then, is the answer to this Nuclear Nightmare? Simply to disarm overnight, to cut back completely on the entire nuclear energy research program, would be naively premature and would be playing right into our enemies' hands—hands that would not hesitate to take immediate advantage of our weakened position.

Already, pro-Communist sympathizers and their liberal altar boys are panting to have us drop our nuclear guard, dismantle the entire American nuclear arsenal and throw ourselves on the mercy of an enemy who is as ruthless as he is cunning.

Furthermore, what are we going to do for power—I mean electrical power? We need vast quantities of fuel (energy) in order to produce the enormous amounts of electricity used daily by our society. Electrical shortages have become critical; brown-outs are certain to lead to blackouts, many of which have already occurred throughout our nation.

There is one further problem. What do we do with the hundreds of thousands of workers in the atomic energy industry, and allied industries, whose bread and butter (and car and house payments) will be cut off when they are out of work?

Is there a way out of this mess? Is there at least a hopeful *approach* toward a way out of this mess? I sincerely believe there is, and the remainder of these pages will attempt to point a way out.

But first allow me to digress for a moment. The story I am about to tell, the facts and concepts I shall present, will sound at first hearing completely incredible. I have always felt that the true art of a non-fiction writer is to make the incredible *credible in time!* The world, to paraphrase H. G. Wells, is indeed running a frantic race between education and catastrophe. New concepts *in new fields of research* are generally opposed by contemporary scientists. Before the advent of the electric-light bulb, no one who did not actually work on the problems solved by Thomas Edison was competent to judge Edison's discoveries. The American Academy of Candlemakers (had there been such an institution) was not qualified to competently judge the invention of the electric light. The Wright brothers' airplane could not be rationally judged by hot-air-balloon enthusiasts. And so, with this brief introduction, let us open the curtain to new vistas that might, just might, save a floundering planet from its present Nuclear Nightmare.

On December 30, 1940, a little-known physician and scientist living in Forest Hills, Long Island, N. Y., composed a letter which he mailed to Professor Albert Einstein, who was then working at the Institute for Advanced Study in Princeton, N. J.

The physician-scientist was Dr. Wilhelm Reich, who was at that time teaching Experimental and Clinical Psychology at New York's New School for Social Research.

Reich, in this brief, important letter,* expressed his hope that he might interest Einstein in his discovery of a specific biological energy which he termed "Orgone" or Life Energy. He had been able to scientifically objectify this Orgone energy thermically, electroscopically, microscopically, and visually, as well as in other ways. He had reached a point in his research, however, where he felt that the magnitude of his discovery, as well as the responsibilities involved, was far too great for his private resources. The discovery of a specific biological energy, which could be demonstrated as *existing in the atmosphere everywhere* as cosmic, primordial energy demanded far wider research scope than Reich could at that time provide. Would Professor Einstein be interested in discussing some of the basic phenomena which Reich had uncovered? In this initial communication, Reich pointed out that he was concerned about the irrational reactions which the scientific community had so often displayed toward new discoveries in the past.

Would Professor Einstein be willing to listen to this researcher and, perhaps, at some later time, visit Reich in his own laboratory to witness additional phenomena at first hand?

Reich was jubilant when he received a short reply from Einstein dated January 6, 1941, which eventually led to an appointment proposed by Einstein for January 13 at 4 P.M. at the Professor's quarters on Mercer Street in Princeton.

Einstein received Reich at about 3:30 P.M. on the appointed day. In preparation for this historic meeting, Reich tested and adjusted an optical device which he had invented to demonstrate

*Photostats of the letters exchanged between Reich and Einstein appear in the publication *Wilhelm Reich, Biographical Material: History of the Discovery of the Life Energy, Documentary Volume A = XI = E, The Einstein Affair,* Orgone Institute Press, Orgonon, Rangeley, Maine, 1953.

visually the existence of the Orgone. He termed this optical instrument an "Orgonoscope." The discussion between Reich and Einstein *lasted for nearly five hours,* during which time Reich introduced Einstein to new facts and concepts relating to the existence of a mass-free, pre-atomic, primordial energy—the same energy which operates in the living organism as the biological or Life Energy. Einstein reacted at first with amazed incredulity. We can only conjecture at the excitement and wonder that must have filled that room as the two men, conversing in German, exchanged questions and answers. Reich was a consummate instructor who had, over a period of many years, mastered the difficulties presented by the totally new field he had discovered. He had so much to tell to the elderly Einstein, whom he respected as humanitarian and first-rank scientist. Reich knew that it would be most difficult to orient a classical scientist in a completely different realm of discovery, that there was "too much" to tell. Over a period of years, for example, he had been able to demonstrate that he could accumulate this newly discovered energy within a very simple enclosure, and that immediately above this "Orgone Energy Accumulator" one could read a consistently *higher* temperature than was evident in the surrounding atmosphere. This finding alone contradicted the second law of thermodynamics—a pillar of classical physics.

The men turned out the lights and drew the blinds, and after they had allowed time for their eyes to become accustomed to the darkness, Reich gave Einstein the Orgonoscope and asked him to look through the instrument and observe the scintillations of the Orgone energy in the room. Einstein was amazed.

Later, Reich discussed the phenomenon of the temperature difference between the Orgone Energy Accumulator and the ambient atmosphere. If it was true, Einstein exclaimed, it would be "a bomb in physics!"

Einstein then agreed that if the facts could be proved, as Reich had outlined them, he would aid Reich and his discovery in every way he could. Reich expressed his hopes that Einstein would *duplicate exactly* the experimental setup which was necessary, and told him further that an Orgone Energy Accumulator would be built and delivered to Einstein's residence. Should any discrepancy in findings occur, Reich stressed that Einstein was to check with him *before* forming any firm conclusion as to the validity of his discovery. Experience and many years of hard-won facts had taught Reich that all too often scientists, when confronted with his findings, had preferred to "explain it all away as being just this or just that," rather than duplicate the painstaking processes that were involved. Einstein agreed to Reich's proposals, and Reich bade the elder scientist a pleasant goodbye. The Orgonoscope remained with Einstein so that he might continue to observe the atmospheric Orgone at his leisure.

About two weeks later, Reich personally delivered a small Orgone Energy Accumulator to Einstein. The apparatus was placed on a table in the cellar. One thermometer was placed in a tube just above the accumulator, while a second, control thermometer was hung some 3½ feet away in the free cellar atmosphere. Einstein then *confirmed the temperature difference.* He requested that Reich leave the accumulator with him for further observations, saying that he would write Reich in a few weeks and discuss his findings.

True to his word, Einstein again wrote Reich, on February 7, 1941. In this letter Einstein mentioned an "objection" which his "assistant" had pointed out to him. The objection came in the form of an "explanation" which this unknown assistant had put forth and which Einstein had uncritically accepted. Einstein thereupon concluded that his assistant's "explanation" thoroughly described the thermal differences noted. And there the matter rested.

Reich was deeply upset by Einstein's rather abrupt response, so different from the cordiality and enthusiasm he had previously received in their personal meetings. He had expected that Einstein would maintain their plan of pursuing any discrepancies of fact to their mutual satisfaction.

After considering the response for nearly two weeks, Reich drafted a lengthy reply to Einstein, reviewing their mutual agreement to pursue any disagreements which might arise. In this response, dated February 20, 1941, Reich wrote a masterful rebuttal which thoroughly dispelled the "explanation" raised by Einstein's assistant, and afforded additional experimental conditions and data for testing the accumulator. Furthermore, Reich cautiously mentioned the biological effect which the accumulator produced on malignant tumors in mice, and outlined his painstaking research with these laboratory animals, stressing the *experimental,* although positive, results of his work.

Again, Reich asked Einstein to consider the immensity of the discovery, the benefits it might render to mankind, and entreated him not to reject this vital new field because of a single "explanation" offered by his unnamed assistant.

Months passed and no word was received from Einstein. Reich could not understand this silence. In person, Einstein had treated Reich with genuine and enthusiastic friendliness. He had grasped difficult concepts immediately and with clarity. He had *confirmed* the thermal difference ("It would be a bomb!"). He understood at once that Wilhelm Reich was neither a fool nor a charlatan. The discovery of a primordial, pre-atomic energy replaced the theory of "empty space" with a tangible, useable, workable mass-free cosmic energy. Einstein had explicitly agreed that such a discovery should be pursued—especially in view of Reich's many long years of personal verification which he and scientific co-workers had established. Now silence!

Reich drafted several letters to Einstein which he never sent. He refused to "push" or to "sell" his discovery to anyone. It would stand on its own scientific foundations, with or without Einstein's cooperation. But Reich could not understand the strange smog of silence that emanated from 112 Mercer Street in Princeton, N. J.

On November 14—*ten months after their initial meeting*—Einstein returned the Orgone Energy Accumulator, but failed to return the Orgonoscope which Reich had left. After *four* letters had been sent to Einstein requesting the return of the Orgonoscope, a note from Einstein's secretary was received on January 16, 1942, advising that the Orgonoscope was being returned.

For many years after this disturbing experience Reich could

find no rational explanation for Einstein's behavior. The discovery of the Orgone was indeed a revolutionary scientific occurrence which threatened to overturn many a pet theory in every area of science. Reich's work had evoked irrational reactions from the Black fascists in Germany and the Red fascists of Scandinavia.

On March 17, 1950, the following article appeared in the *New York Times*:

> **SCIENTIST'S LEAVE STUDIED**
>
> **Einstein Ex-Associate Seeks to Teach Again in Poland**
>
> Ottawa, March 16—Opposition leader, George Drew, in the House of Commons this afternoon raised the question of the propriety of permitting Dr. Leopold Infeld, a former associate of Dr. Albert Einstein and at present a teacher of mathematics at the University of Toronto, to return to Warsaw to organize certain educational programs in cooperation with the Communist Government of Poland.
>
> Mr. Drew said that Dr. Infeld, who had been given hospitality as a refugee in Canada and the United States, had gained considerable knowledge of the latest discoveries in the atomic field.
>
> He had several times stated that he would return to Europe if and when a "progressive government was established in Poland," and he returned there last summer and taught in the Universities of Warsaw and Cracow, Mr. Drew said. Now, Mr. Drew added, he had applied for a sabbatical year during which he would receive half pay from Toronto University in order to permit him to work at educational organization under the Polish Government.

Coming events have a most interesting way of casting their ominous shadows upon our future paths. In this case the shadow had a definite red tinge to it!

When Reich read this article concerning Einstein's assistant, Leopold Infeld, a Communist mathematician, he was convinced that the Red fascists had once again succeeded in obstructing his work. Reich was an avid anti-Communist who characterized Communism as "the organized emotional plague" in his published writings. Many of Reich's associates could not believe that there was a politically motivated move afoot to discredit Reich and to steal his discoveries. It took many years before the pieces of this international jigsaw puzzle began falling into place.

In order to understand the puzzle, we must first learn something about the scientific genius, Wilhelm Reich.

7 Who Was Wilhelm Reich?

The discoveries, harassment, trial, and final silencing of Wilhelm Reich stretch back over some three decades. Let us trace the highlights in the life and career of this brilliant scientist and thinker. In 1968, at the request of author Brad Steiger, I wrote a chapter, "The Silencing of Wilhelm Reich," which appeared in the book *The Allende Letters,* by Steiger and Joan Whritenour, published that year by Award Books. Most of the following data are taken from that chapter, with the permission of Mr. Steiger.

Wilhelm Reich was born March 24, 1897, in imperial Austria. Until 1938 he was an Austrian citizen. Educated first as a private student, he attended a German high school and then entered the University of Vienna in 1918. As a war veteran (he was in the Austrian army from 1915 to 1918) Reich was able to complete the six-year course for his medical degree in only four years, graduating in July 1922.

He did postgraduate work in internal medicine and neuropsychiatry. And while still in medical school, he attained membership in the Vienna Psychoanalytic Society under Professor Sigmund Freud.

Reich was early acknowledged as a brilliant new light on the psychoanalytic horizon. For six years he was first clinical assistant at Freud's Psychoanalytic Polyclinic and during 1924 through 1930 he was the Director of the Seminar for Psychoanalytic Therapy at that institution. Even in those early years, Reich was known for his enormous vitality and scientific integrity. His capacity for work was overwhelming. Indeed, it has often been said of Reich that "he discovered too much."

Besides doing research into the social causation of the neurosis at the Polyclinic, he founded and directed mental hygiene consultation centers in various Vienna districts (1928-30) and in Berlin (1930-33). During 1934-39 he lectured and did research in Orgone Biophysics at the Psychological Institute of the University of Oslo, Norway.

Reich transferred his scientific laboratory to Forest Hills, N. Y., and came to the United States in August 1939. (Later he became a United States citizen.)

From 1939 to 1941, Reich was Associate Professor of Medical Psychology at the New School for Social Research in New York City (the period of the "Einstein Affair") and lectured on his emerging discoveries of the biological and cosmic Orgone energy which he had discovered.*

In 1942 Reich founded the Orgone Institute, which was settled on a 280-acre estate acquired in Rangeley, Maine. The home for the new science of Orgonomy was appropriately named Orgonon. The Wilhelm Reich Foundation was founded in Maine in 1949 by students and friends of Reich (many of them physicians and scientists) to preserve Reich's work and safeguard his discovery of the primordial, mass-free Orgone energy.

Unless one comprehends the pervasive emotional disease of mankind, which Reich discovered and called "the emotional plague" (i.e., *"neurotic, destructive irrationalism at work on the social scene")*, it will be very difficult for the reader to understand why Reich's vital discoveries were, and still are, relentlessly opposed. Nevertheless, some relevant facts follow.

As a young psychoanalyst, Reich searched for the energy (Freud called it "libido") behind the neurotic behavior of his patients. What is it, he asked, that *moves* a patient to feel and to express emotion? Through experimental verification, Reich discovered a bioelectrical charge of energy at the skin surface of the human being during periods of pleasure; and he noted a diminution or absence of this charge during anxiety. (See his book *The Function of the Orgasm.*)

Later experimentation convinced Reich that the biological energy was *not electricity* or electrical energy, but *a totally different energy responsible for Life*! This specific biological, organismic energy he named Orgone, and considered it the *Life Energy per se*!

Through years of careful investigation, supported entirely out of his own pocket, Reich was able to demonstrate the existence of the Orgone in many ways: as energy vesicles in "bions"

(1936-39); in sand (1939); in the atmosphere (1940). In 1940 he was also able to concentrate the Orgone energy in his invention, the Orgone Energy Accumulator.

Reich was able to demonstrate the existence of atmospheric Orgone energy visually, thermically, electroscopically, by way of his field meter, and with a Geiger counter. All of his pioneering experiments were carefully done and were published in books and periodicals, quietly and without the usual fanfare attendant upon such significant discoveries. The quality, scope, and implications of Reich's work were truly revolutionary in the fields of science, medicine, and sociology.

Lest the reader assume that Reich was an isolated crank whose work was neither corroborated nor used by other scientists, the following should be noted. During the period of Reich's greatest scientific productivity, more than a score of top-rank medical doctors and scientists in the United States and abroad published verification of Reich's discoveries in many scientific bulletins and journals.* Thus, quietly and carefully, over nearly three decades, Reich and co-workers investigated the fruitful field of Orgone energy, the primordial, pre-atomic energy of our universe.

In 1950, with the advent of the Korean war, Reich prepared his well-equipped laboratory in Maine to help in the U.S. war effort if needed. At this time he worked out his famous "Oranur Experiment" in which he investigated the possibilities of *an anti-nuclear effect that the Orgone might have on nuclear energy.* Later, Reich charged that the successful results of his Oranur Experiment were "the acute object of espionage activities directed from Moscow. The conspiracy was directed to steal the discoveries for the U.S.S.R and to kill it in the United States." (See "Table of Events" on Public Court Record R. 41-46 and "Appendix of Appeal Brief of Wilhelm Reich, October 1956"—especially "Historical Record of Information Given Regarding Oranur.")

At this point let us pause briefly. Having been personally involved in Orgonomy for nearly 20 years, and having duplicated

---

*See Appendix C.

many of Reich's basic experiments, I am nevertheless still amazed at the "incredibility" of this man's work. I realize, therefore, what these bald facts must look like to the average reader who has no personal experience with Orgone energy. Here was a man —*one man*—who had done pioneering work in several major fields of research. He had made major discoveries in emotional diseases. He had *solved* the riddle of the cancer cell! That was enough for one man—for any man in any one lifetime. But Wilhelm Reich was not an ordinary man, though he was a very human, very *humane,* and thoroughly dedicated research scientist. He went on to discover a primordial, pre-atomic energy, demonstrated its existence *objectively,* and invented instruments to make it *perform work!* Think of what would happen to Mr. John Doe if he invented an automobile engine that could run on *water!* This planet is presently controlled by *power politics.* Without power there can be no politically active and effective control. Our earth, on the sociopolitical scene, is run by power. The possibility of an automobile engine that runs on water would threaten the entire power structure. Every country in the world would desire to have such an engine—just as every major country today is striving for nuclear power.

What effect, therefore, would the discovery of a *free source of cosmic power* have upon the power politics of this planet?

Ralph de Toledano's book *The Greatest Plot in History* delineates the acute espionage activities that resulted in the Communist theft of American atomic secrets. Mr. de Toledano writes, "From the moment that Dr. Einstein wrote to President Roosevelt, the Center in Moscow knew that the United States had embarked on the greatest scientific adventure since the beginning of time." It would be irresponsibly naive to believe the Kremlin was frantically interested in obtaining America's atomic secrets and not at least be equally interested in stealing Reich's discoveries regarding the Orgone energy, *the substratum from which nuclear energy is derived!*

Furthermore, it is no well-kept secret that the major governments of the world (particularly the U.S. and the U.S.S.R.) have been engaged in space exploration in order to make contact with intelligent life on other planets. Despite official denials to the contrary, the UFOs which have been intruding into our earth's atmosphere in ever-increasing numbers are posing major problems.

Reich's discovery of a cosmic Orgone energy *holds the key to the propulsive power of these outer-space craft!* Would such a discovery, therefore, be a "target" for foreign espionage activities?

Now let us return to a review of Reich's work. The 1950 Oranur Experiment led Reich to the discovery of certain highly noxious "DOR" clouds (clouds containing *D*eadly *OR*gone energy) which came over his Maine laboratory repeatedly. He believed these DOR clouds were responsible for widespread planetary drought and desert development. He invented a "Cloud Buster" by which he repeatedly was able to clear the atmosphere of these deadly clouds. The Cloud Buster utilized the principles Reich uncovered during his investigations of the Orgone energy; and with this Cloud Buster, Reich was successful in producing and in stopping rain.

During this entire period of vitally important research, Reich was under almost continuous harassment and attack from snipers. He was constantly being followed. Bugging devices were discovered in his automobile. The air was let out of his tires. His lawn was repeatedly strewn with garbage. He began receiving crank telephone calls. And then came the infamous trial that would end with his death in a Federal penitentiary in 1957.

8 *End of a Titan*

From the long viewpoint of human and planetary history, the case against Wilhelm Reich and Orgonomy (the study of Orgone energy) was undoubtedly the greatest single disaster for the forces of good on this earth. Should this planet survive another century, students of medicine and law in the year 2072 will be as thoroughly familiar with the facts surrounding this incredible struggle of a single, magnificent human being standing, as it were, out in the open and being bludgeoned by enemies who remain to this day largely in the shadows of anonymity.

Reich was convinced that the bloody hands that were attempting to strangle him into silence and steal his secrets stretched across the seas from the Kremlin. My own belief is that Reich was absolutely correct in his charge of a Communist conspiracy against him. Reich had been a prime target for Communist attacks in Norway years before. Orgonomy, by its very nature, is abhorrent to both the Red and the Black pestilence of fascism. Reich's monumental sociological work, *The Mass Psychology of Fascism,* was ordered destroyed in Hitler's Nazi Germany, and Reich himself was forced to flee that country. His book *The Sexual Revolution,* a work that reveals basic flaws in the Communist ideology, drew volleys of fire from pro-Communist and other leftist elements in the United States.

An excellent and incisive work dealing with earlier aspects of pro-Communist attacks against Reich is the book *Emotional Plague Versus Orgone Biophysics,* subtitled *The 1947 Campaign,* by Theodore P. Wolfe, M. D., published by the Orgone Institute Press, New York, 1948. A detailed account of the trial proceedings against Reich may be found in the *Journal of Orgonomy,* vol. 1, nos. 1 and 2 (Nov. 1967), published by Orgonomic Publications, Inc., Box 476, Ansonia Station, New York, N. Y. 10023.

Now, let us look at some of the highlights of the formal case against Wilhelm Reich.

In an injunction dated March 19, 1954, and signed by John D. Clifford, Jr., U.S. District Judge for the District of Maine, the U.S. Pure Food and Drug Administration claimed that Orgone energy did not exist and that Reich's books and scientific literature constituted "misbranding" when associated with his Orgone Energy Accumulator.

In brief, the FDA injunction* implied that Reich was little more than a quack, that he claimed he could cure all kinds of diseases from cancer to the common cold and that the public must be protected from Reich's nefarious schemes.

Even the most cursory reading of Reich's published works, however, proves these implications to be entirely false. Many of the books listed as "misbranding" in the FDA injunction have nothing whatsoever to do with Orgone energy; some were written *years before* Reich's discovery of the cosmic Orgone energy. Among the many publications listed (and subsequently destroyed) by the FDA as dealing with the "cure, mitigation, prevention or treatment of disease conditions" were the following books written by Wilhelm Reich—all of which, incidentally, I have owned and studied for many years: *The Mass Psychology of Fascism; The Sexual Revolution; Character Analysis; Cosmic Superimposition; Ether, God and Devil;* and *The Murder of Christ.* Not one of these books claims to cure a single thing—except, perhaps, ignorance! Nevertheless, on the basis of FDA allegations, which Judge Clifford never questioned (why should a judge believe that FDA agents are anything other than "honorable men"?), thousands of copies of Reich's vital books, journals, bulletins, etc., went up in flames between June and August of 1956!

Reich explicitly refused to obey the FDA injunction which in effect barred him from disseminating any further findings on the functions of Orgone energy—the same energy which Dr. Albert Einstein had seen at his home in Princeton, N. J.; the same energy which had demonstrated its effectiveness in dissolving tumors in cancer-ridden laboratory mice; the same energy which had nullified the deadly effects of nuclear energy in the Oranur Experiment; the same energy which Reich utilized publicly and openly to bring and to stop rainfall; the same energy which Reich

*See Appendix B.

had hooked up to run a motor. This cosmic Orgone energy, according to the FDA, *simply did not exist!*

Reich considered the FDA injunction entirely unlawful, inasmuch as it was "obtained by fraud and deceit" and was obviously contrary to truth and fact. He asserted that his vital research was too important to be stopped by procedures that had no basis in scientific fact. In his "Response" to the complaint, dated February 25, 1954, Reich wrote:

> . . . Scientific matters cannot possibly ever be decided upon in court. They can only be clarified by prolonged, faithful, bona fide observations in friendly exchange of opinion, never by litigation. The sole purpose of the complainant is to entangle orgonomic basic research in endless, costly legal procedures a la Panmunjom, which will accomplish exactly NOTHING rational or useful to human society. . . . If painstakingly elaborated and published scientific findings over a period of 30 years could not convince this administration, or will not be able to convince any other social administration of the true nature of the discovery of the Life Energy, no litigation in any court anywhere will ever help to do so.

Reich returned to minding his own business—which was basic natural research. The FDA, however, persisted, and finally won the case by default, inasmuch as Reich refused to appear in court as a "defendant" in matters about which the FDA knew nothing.

From the very first, Reich's position was that of an eminent and responsible scientific researcher who believed that matters of science belong in a laboratory, not a courtroom. What has the U.S. Pure Food and Drug Administration, which rides herd on America's food, drug, and cosmetic industries, got to do with *basic natural research?* This was one question Reich repeatedly asked the courts. At no time, either before, during, or after the trial, did the FDA produce any scientific evidence that contradicted the voluminous findings of Reich and his co-workers. Reich claimed that the FDA injunction itself demonstrated the lack of responsibility and integrity of those who sought to steal his discoveries and to silence him.

Despite the FDA injunction which ordered him to do so, Reich refused the FDA agents access to his research files and notes. (If Orgone energy *does not exist,* why the desire to read Reich's research files and notes?) Reich refused also to reveal his anti-gravity work. Brought to the courtroom in chains (a totally unnecessary procedure aimed at humiliating a valiant soul), Reich's sole purpose thereafter was "to get the total infamy on the Court Records."

Undoubtedly, a congressional investigation into the FDA's prosecution of Wilhelm Reich will uncover an unprecedented panorama of intrigue and irregularities. For example, Peter Mills, who was the prosecuting attorney for the FDA, *was originally the attorney for the Wilhelm Reich Foundation, for the Orgone Institute, and for Reich personally!*

FDA prosecutor Peter Mills was Reich's own attorney at the time the Wilhelm Reich Foundation was incorporated (Mills actually drew up the incorporation papers). And it was again Peter Mills who notarized the papers *attesting to the motive force* of Orgone energy, which Reich had hooked up to run a motor.*

Thus, according to Mills' own testimony, he was the incorporating counsel for the Wilhelm Reich Foundation and acted as the foundation's attorney from 1949 to 1952. Mills had direct knowledge of, and access to, many of the foundation's highly secret documents. It is nothing short of astonishing, therefore, that this same Peter Mills severed his affiliation with the Wilhelm Reich Foundation in 1952 and in that same year took up employment as a prosecuting attorney for the Pure Food and Drug Administration!

Furthermore, Reich contended in court that FDA Agent Joseph Maguire committed perjury in his testimony. Reich's accusation of perjury concerns Maguire's *denial* that he knew the contents of a volume (volume 41) detailing a Communist conspiracy to steal Reich's discoveries and silence the discoverer.

*See testimony of Peter Mills, R. 348-354, Reich's Petition for a Writ of Certiorari to the U.S. Court of Appeals, for the First Circuit—Supreme Court of the United States, October Term 1956, No. 688.

The FDA's own witness, Ilse Ollendorff, testified that Maguire had been in possession of the "conspiracy" volume. This charge against Maguire is further documented in the book *Wilhelm Reich: A Personal Biography,* by Ilse Ollendorff Reich, published in 1969 (St. Martin's Press).

Thus, Reich asserted, Maguire knew of those who were involved in the conspiracy to steal his discoveries for a foreign power. Reich claimed further that Maguire knew of the search for information at Reich's laboratory, and that Maguire also knew of the intrusion at Orgonon, Maine, by "tools of an espionage ring operating in the U.S.A."

In addition, Reich maintained that Agent Maguire knew about the information relating to the Orgone energy motor and of the sudden disappearance of one of Reich's co-workers, William Washington, who had worked with Reich on the motor and who *vanished behind the iron curtain with two of the motors!*

Yet, Reich pointed out to the court, Maguire never once mentioned his knowledge of the conspiracy to steal these vital secrets; in fact, he *denied* such knowledge, and did so under oath. The court records speak a revealing and unequivocal language.

Reich fought his enemies as far as the law would permit. The U.S. Supreme Court refused to review the case.

One of the most unusual and important aspects of the Reich case brought out during the trial was the research Reich had done with respect to UFOs. On at least two known occasions, Reich had turned his cloud-busting machine on silent, hovering star-like objects in the night sky. My friend Elsworth F. Baker, M.D.,* a long-time associate and co-worker of Reich and editor of the *Journal of Orgonomy,* describes one of these events in volume 1 (1967) of this journal as follows:

> At about 9:00 P.M. one evening, Reich pointed to a very bright star in the southwest, about 40 degrees above the horizon. It was unusual in that it was brighter than I had ever seen Venus

*Dr. Baker maintains a private practice as a psychiatrist and medical orgonomist. He is author of *Man in the Trap* (Macmillan, 1967).

to be. Also, it alternated in color regularly in a sequence of red, blue, green, orange, yellow and white—then back to red. We watched it for several minutes and then observed it through a 3-inch refracting telescope where it appeared larger, but still merely as a body of light. Lining it up with stationary objects, we found that it was moving slowly in a northeasterly direction and changing its position gradually with relation to other stars. Reich then trained the cloudbuster on it, and, after three or four minutes, the red color no longer appeared. This was followed by the successive non-appearance of the blue, green, orange, and yellow colors Now all that remained was the white, which grew dimmer and dimmer until we could no longer see it.

Following are a few paragraphs from Reich's petition for a Writ of Certiorari (On Trial Record under secret code *OROP Desert Ea*):

I [Wilhelm Reich] made actual contact by way of the cloudbuster with luminous objects in the sky on May 12, 1954, between 9:40 and 10:45 p.m. During this hour men on earth saw for the first time in the history of man and his science, *two "Stars" to the west fade out several times* when cosmic energy was drawn from them. The shock of this experience was great enough not to repeat such action until October 10, 1954. The reason for the hesitation was obviously the risk to precipitate an interplanetary war by such experimentation. . . .

The [AAF] Ruppelt report on UFO's clearly reveals the helplessness of mechanistic methods in coming to grips with the problems posed by the spacemen. The cosmic orgone energy which these living beings are using in their technology is beyond the grasp of mechanistic science, since cosmic laws of functioning are not mechanical but what I term "functional." Even the mathematical theory necessary to formulate these problems and make them technologically useable cannot use any of the old mechanistic methods of thought to cope with the functional orgonomic facts.

The helplessness of mechanistic thinking appears in the tragic shortcomings of our fastest jet fighter planes to make and hold contact with UFOs. Being unavoidably outdistanced is not flattering for military pride. The conclusion seems correct: Mechanistic methods of locomotion must be counted out in coping with the spaceship problem. Neither propeller nor jet will or can get us into space beyond.

At the time that Reich was fighting for his life, some 15 years ago, the entire subject of UFOs was "officially" ridiculed. Not so today. One has only to peruse the Hearings conducted in 1968 before the Congressional Committee on Science and Astronautics. This document, published by the U.S. Government Printing Office, titled *Symposium on Unidentified Flying Objects,* shows clearly that Wilhelm Reich was again far ahead of his time in understanding the reality and urgency related to the intrusion into our skies of spacecraft from other planets. Yet continuous rumors were being spread that Reich's interest in UFOs clearly indicated that he was "insane"! A number of highly qualified and highly placed scientists appeared before the Congressional Committee on Science and Astronautics; and many of them voiced concern over the UFO problem, indicating that our Government should pursue the subject thoroughly. A recent issue of *Saga* magazine (Feb. 1971) carried a feature article "A Nuclear Physicist Says UFOs Are Alien Spacecraft"! The subject of the article, Stanton Friedman, is a scientist "with 14 years experience in such highly specialized fields as development of nuclear aircraft, nuclear power plants for space application, and fusion rocket research." Friedman, according to the article, "is convinced on the basis of 11 years of research into this taboo subject, that the origin of UFOs is extraterrestrial, that the Air Force is sweeping the real facts under the carpet, and that Dr. Edward Condon should have retired at 65 instead of heading the deficient Air Force UFO study in Colorado at age 66." This article by B. Ann Slate and Fritz Kron then goes on to detail specific UFO sightings which continue to be "covered up" or "explained away" by officialdom.

Reich's eminent position as a researcher and his flawless reputation for scientific integrity give rise to some highly important questions. Why did a man of Reich's brilliance and stature refuse to comply with the FDA injunction? According to Dr. E. F. Baker's article "Wilhelm Reich" (*Journal of Orgonomy,* Nov. 1967),

The attack [on Reich] in 1947 was initiated by Mildred Brady, whom Reich had found to be sympathetic to the Soviet Union. Significantly, during the period of the trial, *the FDA itself came under investigation and a number of its employees were discharged as Communists.* Orgonomy and Communism are inimical. For a long time, Reich had been quite vocal in effectively exposing and denouncing the mechanisms of Red Fascism. *The Soviet Embassy had ordered all of Reich's books,* and Dr. Walter Hoppe of Tel Aviv reported treating a Russian *who stated they had accumulators in some Russian hospitals.* [Italics added]

Now we have a good idea why Reich repeatedly refused FDA agents any access to his notes and records. In one instance, Reich posted armed guards at Orgonon to repel trespassers, which included FDA agents.

It should also be carefully noted here that Reich was neither a fool nor insane. A court-directed psychiatric examination adjudged Reich perfectly sane. Thus, any shadow of insanity must perforce fall elsewhere.

In his publication *CORE,* chapter 2, "Space Ships and Desert Development," Reich made some astounding assertions. He believed that UFOs are utilizing the cosmic Orgone energy to propel themselves, and that the spacemen are thoroughly familiar with Orgone energy functions. He believed further that people from outer space may have decided to investigate the irritation of the atmospheric Orgone energy caused by our planet's repeated testing of nuclear devices.

Reich also made this grave observation:

> It is not rational to assume that wars are always fought only the way we know them. . . . We must permit ourselves to think of war as a silent, unnoticed slow-working but deliberate destruction of life on a planet, a satellite, or even a star. We must be ready to change every single view we have held sacred if it contradicts *new* facts and *new* experiences.
>
> War may be going on right now with no one being aware of it, with men dying, with trees bending like rubber hoses, green pastures turning into dust bowls, and with academic and civil institutions explaining it all away as with "just this" or "just that." In short, it may turn out correct what one would otherwise feel inclined to ascribe to a schizophrenic

> mind—namely, that instead of shooting *at* the victims of war with bullets, one could very well sap life energy *out of the war victims* with machines which operate according to the orgonomic potential of the Cosmic Energy.

Reich believed that "our planet is in deep trouble." It was this belief that drove him to fight so courageously to protect his Life-positive discoveries and to get the facts into the public record of the court trials.

Thus, mankind is left with Reich's great legacy of discovery and work, and with his grim warning regarding the continuing intrusion of spacecraft into our atmosphere:

> We are facing an emergency, facing it not only as the species mankind; the principle of LIFE ITSELF on Earth is challenged. We can no longer permit ourselves to be as petty as to continue our Little Men's quibbling over nothing, as if nothing at all has happened. *We are in a process of deep and crucial change of our total existence, biological, physical, emotional and cosmic.* Let us, therefore, start behaving in accordance with these facts.

The courts refused to take Reich's warning seriously. And so, despite the fact that Reich informed the court that he would never survive imprisonment (he had a severe heart condition), he nevertheless was found guilty and was thereupon incarcerated. He died eight months later—*only a week before he was to be released.*

Far from silencing Wilhelm Reich, the FDA injunction and trial, the burning and banning of orgonomic literature, and the untimely death of this great genius have brought worldwide attention to Reich's revolutionary discovery of a pre-atomic, primordial cosmic Life Energy. Also, the relentless desert development of our planet continues unabated, and the threat posed by UFOs can no longer be evaded. Reich's work, specifically his Oranur Experiment, points the way to the "peaceful atom" which has thus far eluded our best scientific minds. His Cloud Buster affords a positive approach to the breaking of nationwide and planetary droughts, and to the recession of worldwide desert development. (One American Cloud Buster, for example, could easily have removed the black cloud that hovered over Apollo 14, and delayed a liftoff for more than 30 minutes while engineers and physicists stood helplessly by. One or more Cloud Busters installed strategically at various locations of the Northwest could quench the conflagrations that decimate our magnificent forests each year.)

No, Wilhelm Reich has not been silenced. As long as there are public court records and courageous men who respect truth and justice, Reich's work will continue to flourish. Truth has merely been delayed. Justice, however slow in coming, must inevitably remove her blindfold.

9 The Orgone Energy Accumulator

A man has no ears for that to which experience has given him no access.
Nietzsche

When I saw my first Orgone Energy Accumulator (hereafter referred to as "oraccu"), I looked at it with undisguised contempt. It didn't look like much to me—a midget-size telephone booth, with a seat inside, but with no electrical connections, no fancy chrome knobs. What good could it do?

That was nearly 20 years ago. Since that time, with continual use over the years, my respect and admiration for this superb invention has steadily grown. The oraccu has saved me thousands of dollars in medical bills, kept me in good health and become such a constant friend to me that I could not visualize an existence without it.

My first oraccu was obtained on prescription by a medical doctor who was a practicing medical orgonomist. Over the years I have built others myself, including oraccu blankets, Orgone energy "shooters" of various design for local application, and oraccu apparatus for experimental work with plants.

This is not meant as a technical discussion of the oraccu. Those who wish experimental data are referred to the literature mentioned in the Bibliography.

Orgone energy is the primordial, mass-free cosmic Life Energy in which we and everything on our planet are immersed. It was first discovered and objectively verified by Dr. Wilhelm Reich between 1936 and 1939 in Norway, with subsequent work being done by Reich and co-workers in the United States.

The oraccu was invented by Reich to collect and accumulate the Life Energy in our atmosphere and make it useable for scientific and medical purposes. The specific color of Orgone energy is blue or bluish gray, although it changes in color and form according to its concentration and state of excitation. Orgone energy is able to penetrate *everything*, but at varying speeds.

Although it is present everywhere, it varies considerably in concentration and quality.

Reich discovered that organic matter (wood, cotton, wool, etc.) readily *absorbs* Orgone energy, whereas metallic matter first absorbs *but then repels* the energy rapidly. Thus was born the fundamental principle governing the construction of the oraccu and all oraccu devices.

The basic oraccu is a simple six-sided box, all walls of which are constructed in exactly the same manner. The *outermost* layer of each wall is constructed of organic material, such as wood or pressed composition board (Celotex). The *inner,* metallic layer can be made of sheet iron or iron wire mesh (screening). Attention should be paid to both the absorbing (organic) and repelling (metallic) layers, as to the amount and type of materials to be used. For medical and scientific purposes, Reich recommended *iron* as the best metallic material. Other metals, for example aluminum, he considered harmful for medical purposes. Organic materials vary also in their ability to absorb energy as well as moisture.

Orgone energy is strongly attracted to, and absorbed by, water. Therefore, on damp or rainy days the oraccu will not function as well as it does on drier days, as the Orgone is absorbed by atmospheric moisture. In general also, climatic conditions and geographic locations can considerably influence the effectiveness of the oraccu.

The specific layering of organic outer wall and metallic inner wall constitutes a "one-fold" oraccu. I have built and used oraccu devices up to ten folds for experimental purposes on plants; however, Reich warned against using an oraccu stronger than three folds for medical purposes without proper medical supervision.

The layering of organic and metallic walls establishes a *direction of flow* of Orgone energy from the outside to the inside of the oraccu, where the Orgone is concentrated above the level in the surrounding atmosphere. Orgone energy flows *from the weaker potential to the stronger potential* (just the opposite of electrical flow). Thus the human being within the oraccu attracts the energy to his own orgonotic system, which is higher in potential than the oraccu.

Depending upon the individual's specific level of bioenergy, it may take several weeks or even months of regular, daily 30- to 60-minute sittings within an oraccu before subjective sensations are felt. In my case it took a few weeks of daily 30-minute use before I felt sensations of warmth and prickling on my skin surfaces. The less or lighter the clothing one wears in an oraccu, the more immediate the effect, as the energy is first absorbed by clothing before it can penetrate the skin.

The Orgone energy level of the user must first be raised to a level sufficient to interact with the energy of the oraccu. I felt the typical "glowing" of my body, the warmth and relaxation which became familiar signs that "something was happening." The diminution and final cessation of this warm glow indicates that the organism has had "enough"—that nothing further is happening.

At first I used the oraccu with a small electric light (7.5 to 15 watts) to read by. But as I became familiar with my subjective sensations and slowly lost my fear of the Orgone energy, I turned off the light and sat in total darkness, gradually becoming aware of the visual manifestations of this Life Energy of our universe.

For many hundreds of hours I have sat in the "darkness" of the oraccu and watched the streaming, scintillating pulsations playing around my hands, fingertips, face, and body. When I felt "good" (that is, relaxed and expanded), I could make excellent visual contact with the energy. When I felt contracted and anxious, such contact was often difficult and frightening. It should be noted here that I was using my first oraccu under the supervision of a medical orgonomist.

After a year of continuous, daily use, I felt stronger and more alive than ever before. My frequent colds were far less severe; my chronic mastoiditis, which I had had since childhood, disappeared, and I enjoyed excellent health.

My interest in and respect for the oraccu grew with use and experience. I built small oraccus and began repeating Reich's

TO–T experiment, which demonstrates consistently higher temperature immediately above the oraccu. I then built an orgonoscope. For several months I kept pulse and temperature readings, jotting down my pulse rate and temperature five minutes before using the oraccu and then again during the periods of glowing. I found that my temperature rose during these glowing periods as much as 1.5 degrees Fahrenheit, while my pulse rate slowed and became stronger.

I constructed small "shooter" tubes, invented by Reich, using hollow BX cable as the inner metallic layer, wound with alternating layers of cotton and steel wool. These I found very useful for local irradiation of cuts and burns. The simplest first-aid shooter consists of a glass test tube stuffed loosely with steel wool and stoppered with a wad of cotton. Several 30-second irradiations of my throat with the test-tube shooter have successfully routed a beginning sore throat. And daily one-minute irradiations of my teeth and gums have greatly reduced the number of cavities I suffered each year.

As a former U.S. Navy Hospital Corpsman, with more than four years of active duty in World War II and the Korean war, I have had considerable experience with cuts, burns, wounds, and infections—their usual treatment and manner of healing.

I have used my shooters on many personal cuts and burns, with very good results. For burns (including frostbite) I first coat the area with petroleum jelly—to aid in the absorption of energy—and then hold the shooter about ¼ inch from the area for 15 minutes every hour for the first three to five hours, depending upon the severity of the burn. Initially, the pain returns in full force, gradually diminishing with each successive irradiation. After the first day, I continue the irradiation from three to five times per day (15 minutes each time). Rarely have I experienced the pain and discomfort usually attendant upon second- and even third-degree burns. Healing has always been rapid and complete.

Several years ago while driving the Alaska Highway in winter (a journey I would never recommend at that time of the year), I suffered frostbite. The tips of two fingers turned white, with compete loss of feeling. Three days later I began application of the test-tube shooter to the fingertips; after two days of treatment, both the color and the normal feeling returned to the fingers.

As for cuts and bruises, I have found that the sooner they are irradiated with an Orgone energy shooter, the quicker they heal. I apply the usual antiseptic *after* the first 15- to 30-minute irradiation. Then, with dressing in place, I continue irradiating the cut for 15-minute periods, three to five times daily.

Once I lacerated my thumb with a scraping tool while working in a metal shop. The laceration required suturing, and when I returned from the hospital I immediately used the shooter on my thumb. I had done quite a bit of minor surgery in the Navy, but never had I seen a laceration heal so rapidly and cleanly as this one. The sutures were removed after three days!

Local infections, including boils, have always yielded to repeated applications of the shooter. I have also found that *general, total body irradiations of Orgone energy,* by way of the oraccu (box type or blanket), aid immeasurably in building up my resistance to infections.

Also I have found that there is no substitute for regular, daily total body irradiations of Orgone energy as insurance against colds. Often, when I have felt a cold coming on, several prolonged sittings in the oraccu have completely arrested or minimized the cold symptoms. On at least four occasions I went through the entire sequence of cold symptoms in a matter of several hours, while using the orraccu for 30-minute intervals during those hours.

When my wife and I lived in the Alaskan wilderness, miles from the nearest doctor, one of the first things we did was build a wooden shed to house our oraccu. We relied on the oraccu for our continued good health, and we used the shooters as a matter of course on ourselves as well as on our dogs. The dogs were often cut by the thorny devil's-club which grew profusely in our area. And our canine friends were also the victims of broken bottles tossed by careless tourists or hunters.

I have grown a number of plants in small oraccus and

watched them develop faster and more vigorously than control plants. It should be noted that *oraccus must fit the physical dimensions of the user,* with the walls at most only a few inches from the body for best interaction of oraccu and individual.

I once built a cubic-foot oraccu which soon became the personal property of a pet cat, for whom I left the lid open. It was interesting to watch him jump into his oraccu whenever he felt the need, and use it for varying periods, sitting and yawning inside, with his great glowing eyes fascinatedly following the energy.

My ten-year-old German shepherd often shared the oraccu blanket I made, burrowing under the blanket with his nose until I let him lie beside me. The oraccu blanket is simple to construct. A large wool blanket (some synthetic blankets are also suitable) is layered with galvanized *iron* wire mesh, the kind commonly used for window screening. This is followed by a sheet of plastic sheeting (used in the building trade as a vapor barrier), then another sheet of screening. The blanket is easily sewn together around the edges with heavy thread or can be stapled with a heavy-duty stapling gun. I usually make my blankets approximately three by six feet, using one to lie upon and the other as a cover. The blanket has several advantages over the box-type oraccu. It is far lighter and can be transported with ease; it can be aired out with little difficulty, and it is less expensive to construct.

Nearly two decades of using the oraccu have taught me some valuable lessons. The oraccu operates *functionally,* not mechanically, and responds to variations in weather in the same manner as does the living organism. On raw, "contracted" days the oraccu feels similarly contracted, and longer periods are required before I respond to it. On clear, sparkling days (which seem to be getting scarcer on this planet) the oraccu feels stronger and more immediately responsive.

Since Orgone energy operates in both the atmosphere and the living organism, whatever has an adverse or Life-inimical effect upon the living will similarly change the quality of the Orgone energy in the oraccu. High-voltage equipment, radioactive materials (including radium-dial watches and clocks), TV sets, fluorescent lights, harmful gases and chemicals, etc., *should never be used within or in the proximity of any oraccu device*! One should not use an oraccu while undergoing X-ray treatments or while taking any medication, without first consulting a medical orgonomist.

The user must always be on guard against his own *contempt* of the oraccu, of overusing it, or in *testing* it. I have known people who never aired out their oraccus, or who kept oraccu devices locked away in stuffy closets for months. In such homes there was always the highly charged heaviness typical of Oranur, which can be very dangerous to Life.

During periods immediately before and after nuclear tests, I have found my oraccu repelling, unuseable, and requiring more than normal ventilation or airing out. The oraccu also reacts strongly to the noxious DOR clouds (clouds containing *D*eadly *OR*gone energy) which appear at irregular intervals over any area, and which Reich associated with UFOs, or outer-space craft, in our atmosphere.

DOR clouds look like dirty, dark-gray-to-black smudges in the sky. They remind me of shredded steel wool, or of flaklike puffs of "blasted clouds" in the sky. They give a feeling of ugliness and dirtiness, are nauseating (literally), and give one the feeling of death. When DOR clouds come over an area, one can see the sparkle go out of the atmosphere and the leaves, trees, and wildlife appear forsaken. The animals, in particular, move sluggishly; birds fly low. And you will see many dead animals on the roads—many more than usual. DOR lacks oxygen and water and will attack plants, rocks, and man to obtain the needed oxygen and water which it consumes from its host.

During periods of Oranur or DOR, it is advisable to "draw off" the highly charged or stale, deadly energy by grounding the oraccu by means of several hollow BX cables which lead from the bottom of the oraccu (or blanket) into a bucket of water. The water should be changed at regular intervals, preferably every few days.

Now I feel that a word of caution is in order for those who

are unfamiliar with the oraccu. The serious student will find his way by working slowly and cautiously, and by first studying every bit of literature that he can. In working with the Orgone energy, we are dealing with an energy many times more powerful than nuclear energy. Orgone energy functions are slow, gradual, often operating in imperceptible stages—as in the slow buildup of clouds or the flowering of a plant. One must strive always to be a patient observer before attempting to take a single new step. The oraccu is neither a miracle-working panacea for the ills of humanity, as some have characterized it, nor a nonsensical piece of charlatanism, as others would have it. It represents a magnificent achievement in natural science which the late Dr. Theodore P. Wolfe called "THE MOST IMPORTANT SINGLE DISCOVERY IN THE HISTORY OF MEDICINE, BAR NONE."

Where the oraccu will lead mankind depends upon those who have the scientific capacity to understand its realm of functional laws, to observe and further its scientific and medical potential, and to protect this discovery from those who would destroy or misuse it out of fear, hatred, or avarice.

Note: After the manuscript for this book was completed, Barbara G. Koopman, M.D., an associate editor of the *Journal of Orgonomy* and a medical orgonomist, was kind enough to review the work. Her comments on this chapter are pertinent; and I thank her for them: "One or two of your observations on the Orac [Orgone Energy Accumulator] need updating: for instance, sitting 30-60 minutes per day is probably too much. (It may be contraindicated entirely in the average urban Dor-laden atmosphere.) Also, when grounding it in water to draw off Dor and excess charge, the water needs to be changed much more frequently than we originally thought."

10 A Question of Credentials

More than twenty highly qualified medical doctors in the United States and abroad have written articles on various aspects dealing with Orgone energy in the treatment of their patients. Many of these articles are landmarks in the history of pioneering research in Orgonomy or, more specifically, in medical Orgonomy. In 1956, tons of this vital research were consumed by flames on orders of FDA agents.

Public opinion is a fickle mistress whose head may be turned one way or the other by an insinuation, by smart alecks who believe that being witty is equivalent to being wise. Malicious rumormongering, slipshod journalism, and society's ever-present hunger for the salacious and the sensational often combine in deadly alliances where new discoveries are concerned.

Great men and women in every age have faced the black quagmires of misunderstanding and ridicule. Socrates, Christ, Giordano Bruno, Galileo, Semmelweiss, Franz Anton Mesmer, Mozart, Van Gogh, Freud, the Curies, etc., etc., were outcasts, each in his or her own way.

It is a mark of our times that a book like *Myra Breckenridge* can sell in the millions of copies, while a book entitled *The Function of the Orgasm,* by Wilhelm Reich, is banned by agents of the FDA! It is a sickening and shameful fact that millions of people will support pornography while simultaneously turning their backs upon vital information that could help them out of their morass. Man (distorted, modern man) simply doesn't know the difference between good and bad—and by that I mean, *healthy and sick*! Ninety-nine times out of 100, when given the choice between clean water and filthy water, the public citizen chooses the filthy water. Or, if he has only clean water, he will spit into the water, or befoul it in other ways.

Unless we come to understand the sharp difference between the healthy and the sick, what will save this planet from total annihilation? These are harsh words, I know; but these are harsh times, *critical* times.

Millions of Americans have been titillated by defamatory articles written about Wilhelm Reich and his "orgone box" in the twentieth century. Millions have died, and millions more will die, out of ignorance of the work of Dr. Reich. Those deaths are useless, futile deaths—and many more deaths will be just as useless, just as purposeless.

Now time is running out—and we are running out of time. All of the witty innuendoes about Reich and his "nutty ideas" giggled about at cocktail parties; all of the malicious insinuations in national stories about Reich's "sex box"; all of the malicious undermining behind closed doors in psychiatric seminars and medical convention rooms; all of the cheap laughter at "putting down" a great man ("Let's See The Orgone Cure a Two-Year Jail Sentence," headlined the New York *Post*)—all of this *garbage* has returned to strangle man in his own pollution!

Drip by dirty drip we have poisoned our wellsprings of Life and creativity. One day we will have to learn to tell the difference between a big-mouth blackguard who is nothing more than a mechanical walkie-talkie (ask him a question—*any* question—and you'll get an opinion!) and the man of knowledge and practical work. One day, as Reich predicted, we will learn to tell the difference between a *killer of Life and a protector of Life by their emotional expression*, just as any good woodsman knows the difference between deadly snakes and good snakes! A man cannot long survive in a jungle, including our "civilized" jungle, unless he knows the difference between simple good and evil, healthy and sick, rational and irrational.

One way to immediately start learning the difference is to make a clear-cut distinction between *fact and truth* as opposed to *opinion about fact and truth*! Society progresses on the bedrock of hard-won factual knowledge. This seems so self-evident as not to be worth repeating. You don't run an airline, a trolley car, a classroom, a courtroom, a business, or a household *on mere opinion*! To do so is sheer, suicidal folly. And right now, at this moment, our entire planet is suffering from a horrendous case of just such suicidal folly.

How do we get out of this suicidal mess? There is no sense

in relating example after example of man's daily folly, unless we can somehow figure out a way to the door leading out of our present chaos! Reich gave us one of the most important bits of insight to aid us through our present crisis: *Demand factual opinions*—that is, conclusions drawn from facts! Erroneous, flippant opinions *can be murder*! Erroneous opinions regarding Dr. Semmelweiss' discovery of antisepsis killed hundreds of thousands of mothers and babies.

In order for an opinion to be rational, the opinion-giver must know at least—*at least*—as much about the subject under discussion as the man he criticizes, or the work he criticizes, or the procedures he criticizes.

Now, let us look at how opinion, unfounded, biased opinion, can destroy a good man and his work. The incident I am about to relate is only one of dozens, possibly hundreds, of similar incidents which, drip by drip, poisoned public opinion with respect to Wilhelm Reich and his discovery of the Orgone. In this instance, however, the walkie-talkie opinion-maker did not achieve what he set out to accomplish.

In December of 1962, my friend Oscar Tropp, M.D., a former Army psychiatrist and a practicing medical orgonomist, suggested that I might want to attend a meeting of the American Humanist Society. The New York chapter of this society was sponsoring a series of lectures on various subjects that month. A particularly interesting lecture was scheduled for December 16. The lecture, "Quack Medical Devices in New York City," was to be given by Dr. Irving Witlin (his doctorate was in law). The announcement mailed out by the Humanist Society stated:

> Are we more civilized today than people were in Mesmer's time? Would members of the "under-developed" nations laugh at our gullibility? Mr. Witlin, administrator of the Education Department, Bureau of Food and Drug, NYC Dept. of Health, will discuss orgone accumulators, different devices to improve sexual potency, arthritis, personality defects and insanity. Mesmer still lives.

I was personally gratified to learn that "Mesmer still lives," inasmuch as I had translated and published the basic works of Dr. Franz Anton Mesmer. My interest in Mesmer coincided with my interest in Reich, because, as I learned through many years of research, Mesmer had stumbled upon concepts and methods quite similar to Reich's discoveries. In fact, it was my knowledge of Orgonomy that had led me to investigate Mesmer's work, where I found that, like Reich, Mesmer had suffered ridicule, exile, and finally heartbreak. His contemporaries slandered him mercilessly, without ever bothering to duplicate his procedures. And although Mesmer never practiced hypnotism a day in his life, to this day he is considered "the father of hypnotism," while his genuinely pioneering and vital therapeutic procedures remain largely unknown! I dealt with that subject in an article, "The Emotional Plague Versus Animal Magnetism," which appeared in volume 1 of the *Journal of Orgonomy*.

So I had two excellent reasons for wanting to attend that lecture, and I thanked Dr. Tropp for informing me of it.

I was the first one at the lecture hall that Sunday evening in Greenwich Village, New York City. About 100 people filled the hall used by the New York chapter of the Humanist Society. Dr. Witlin, a chunky, energetic man with glasses, arrived early with an armload of contraptions and devices which he nervously arranged before him on a table. As he set out each item, he joked informally about each bizarre piece of equipment. These included a plastic, cuplike device which he referred to as a "bust developer." Then he produced an electrified copper plate, and "electro-therapeutic bracelet," priced at $7.50, "to cure arthritis," he said, laughing. This was followed by a "portable electro-dynamometer," which, he said proudly, "we are picking up now" ("we" meaning the Department of Health).

This informal joking period was obviously being used to set the stage and mood of Dr. Witlin's formal lecture.

After being formally introduced by the presiding member of the Humanist chapter, Dr. Witlin referred to his table laden with contraptions, remarking that anyone who cared to might examine them after his talk, which he gave partly extemporaneously, partly by referring to notes.

As soon as Dr. Witlin began his formal lecture I realized that much of what he was saying originated from the book *Fads and Fallacies in the Name of Science,* by Martin Gardner, published in 1957 by Dover Publications. By an odd quirk of fate, I had just spent several days looking into this book, and other books written by Mr. Gardner. I had a strong desire to write a book review of Gardner's *Fads and Fallacies,* and had done my homework thoroughly, at the public library at 42nd Street and 5th Avenue.

Dr. Witlin warmed up to his subject with the usual slanted and defamatory references to Franz Anton Mesmer, such as can be found in a number of the four-bit paperbacks that cram the shelves of secondhand bookshops (reduced for quick sale to two for 15 cents).

The last portion of Dr. Witlin's lecture was devoted exclusively to Wilhelm Reich and the Orgone Energy Accumulator. Throughout his entire lecture, Dr. Witlin used the same tactics employed by Gardner in *Fads and Fallacies*—sly innuendo, guilt by association with the grotesque and the bizarre, and "joking" contempt. Among other things, Dr. Witlin said that "Reich was kicked out by both the Nazis and the Communists"! (And what, I asked myself, is so damning about that?) Then he attempted to connect Reich's work with other "insane racial theories." (A blatant falsehood which no one who has made the slightest serious study of Reich's work could ever voice.)

Among his other distortions and misstatements, Dr. Witlin said:

1—Reich advocated "the full, uninhibited release of the sexual urge." This is the same slanderous pap uttered by pro-Communist sympathizers back in 1947, and picked up again by Witlin from page 250 of Gardner's *Fads and Fallacies,* 2nd edition.

2—"Reich went queer," said Dr. Witlin—a superb remark emanating from a man who has studied law!

3—"Reich discovered *orgonic* energy, which had to be blue in color."

4—Reich "invented a tube which when inserted into the vagina could destroy bacteria." This is a twisted truth, taken out of context to make Reich appear, again, like some "mad sex cultist." What Reich *did* say can be found in a 56-page pamphlet, "The Orgone Energy Accumulator: Its Scientific and Medical Use," which discusses the experimental nature of Orgone energy on specific disease syndromes. In this pamphlet there is *one reference* to the fact that bacteria in the vagina become immobile after being irradiated by a glass pipe filled with steel wool. Reich describes, in the pamphlet, results of Orgone irradiation on the blood cells, tissues, organs, wounds, burns, etcetera. Yet Dr. Witlin digs out *the one single reference made to the genitalia* and distortedly cites this single reference! And they called Reich the leader of a "sex cult"!

5—"Dr. Michael Silvert, a disciple of Reich's, shot himself." Another false statement. Dr. Michael Silvert, a co-defendant in the FDA's case against Orgonomy, was released from prison before Reich, and committed suicide, but not with a bullet.

6—"Reich found a way to separate his disciples from their money." This is the same malicious lie used against Dr. Mesmer 200 years ago. The truth of the matter is quite the opposite. Many physicians were trained by Reich, both here and abroad. The knowledge Reich imparted to these doctors aided them immeasurably in their practices, for which they were justly recompensed. Reich died with very little money. He poured a personal fortune into his research—money which he never saw again.

Dr. Witlin's other statements included a description of the Orgone Energy Accumulator, in which, he claimed, "a person certainly might become asphyxiated, since it was completely closed." (All experimental accumulators had ample air spaces built into them.) Witlin continually referred to the accumulator as a "quack device."

In summation, Dr. Witlin remarked that a quack can often get "venal" though well-trained physicians to become associated with him. Without mentioning his source of information, Witlin used Martin Gardner's self-serving criteria for being able to detect a "paranoid pseudo-scientist."

The beaming Dr. Witlin thereupon completed his lecture by

asking the audience to "make sacrifices," if need be, in backing their local agencies and the Federal Food and Drug Administration.

After a 15-minute intermission, an informal question-and-answer period was begun. A member of the Humanist Society rose in the audience to inform the gathering that the society was now stocking Martin Gardner's book *Fads and Fallacies,* which could be purchased at this hall. (That *was* a surprise!)

A young man got up and informed the audience that he was now engaged in duplicating some of Reich's work and that he realized that much of Reich's later works "reads like science fiction." He said, "As far as I am concerned, Reich's discoveries were just what Reich claimed for them."

The director of a private research group was recognized next. He stated that the case against Reich was far from over, that he himself had verified many of Reich's findings and that the FDA had never produced any evidence to substantiate their contention that the accumulator was worthless.

Dr. Witlin had no comment.

A woman in the audience next corrected Dr. Witlin's statement about asphyxiation in the accumulator, informing him that all accumulators were adequately ventilated.

Dr. Witlin had no comment.

Finally, Dr. Witlin recognized my hand, and the following exchange took place.

Eden: Dr. Witlin, since you are a lawyer, I assume that you make a sharp distinction between fact and opinion.

Witlin: I certainly do.

Eden: In your lecture this evening, you relied rather heavily on Martin Gardner's book *Fads and Fallacies.* I know, because I only recently finished reading his book.

Witlin: I did not rely heavily on Martin Gardner.

Eden: All right. Would you say you relied "slightly" on Martin Gardner?

Witlin: I would say slightly.

Eden: The only reason I mention Martin Gardner is because, first of all, *you* quoted Gardner several times tonight; and, secondly, the Humanists are selling *Fads and Fallacies.* And I would like to make it perfectly clear that Martin Gardner is not an authority on Wilhelm Reich or Orgonomy. *Fads and Fallacies* is not a scientific book, nor is Martin Gardner a scientist. Mr. Gardner wrote a book called *Twelve Tricks With a Borrowed Deck.* [The audience began laughing.] Mr. Gardner is a writer of funny jokes and tricks. He wrote another book called *The Annotated Alice.* It is all about Alice in Wonderland. Now, if you want to read about Alice in Wonderland, read Martin Gardner. But if you want to read about Orgone energy, read Wilhelm Reich! Not Martin Gardner! Gardner is also co-author of a book called *The Wizard of Oz and Who He Was.* [More laughter.] Mr. Gardner also writes a column of monthly games and tricks for the *Scientific American.* So if you want to read games and tricks, read Martin Gardner.

[The chairman of the meeting interrupted me, stating that this was not "a discussion of Martin Gardner."]

Witlin: If you have a question, please ask it.

Eden: I have a question, two questions, which I would like you to answer. The first one is this: Have you, Dr. Witlin—I'm not talking now about the Food and Drug Administration or the Department of Health—*have you yourself, personally, ever used an Orgone Energy Accumulator*?

Witlin: I have not.

Eden: *Have you ever repeated one single experiment given by Dr. Wilhelm Reich in verification of his discovery of Orgone energy—you, yourself, personally*?

Witlin: I have not.

Eden: *Then you have shown yourself to be completely incompetent and unqualified to discuss something about which you yourself know absolutely nothing*! That's all I have to say.

Witlin: I can only say this, that the FDA had a case against Reich, and a jury found him guilty.

Eden: By default, Dr. Witlin! By default! There has never been produced a shred of scientific evidence to contradict anything Dr. Reich did or wrote. *There is a difference between fact and opinion!*

I sat down. There was quite a stir in the audience now. Dr. Witlin recognized the hand of a tall, elderly gentleman seated on the left side of the hall.

"Dr. Witlin," the man began, "Sigmund Freud considered Wilhelm Reich a very great scientist. And when you stand there, and without any evidence you say that 'Wilhelm Reich was *queer*, he was crazy,' where is your evidence? [Several people in the audience repeated the man's question.] When you say a thing like that," the elderly man continued, "I think such a remark is *absolutely lousy*!"

Witlin: I can only repeat that the orgone box was declared worthless and that Reich and Silvert were jailed.

A woman rose to ask Dr. Witlin to substantiate a statement he had made to the effect that Reich claimed he could cure cancer. "Isn't it true," she said, "that Wilhelm Reich never claimed he could cure anything?"

Dr. Witlin had no comment.

The next man Dr. Witlin recognized read a section from Reich's *The Cancer Biopathy* (destroyed on FDA orders) to show that Reich did not claim he could cure cancer, but merely compiled his observations, the failures as well as the successes, and the effects the Orgone Energy Accumulator produced on cancer patients.

Now a man rose to state that although he had come as a skeptic, what he had heard this evening "was very interesting." He asked Dr. Witlin where he might obtain any of Reich's books to read for himself. Dr. Witlin said that he did not know where books could be obtained. The man also wanted to know if there was any legal prohibition against his using an Orgone Energy Accumulator for experimental purposes.

"You mean for research?" asked Dr. Witlin.

"Yes, for research."

Dr. Witlin said that there was no legal prohibition against using the accumulator for research purposes.

"But," the young man continued, "since your department burned Reich's books, where can you get them?"

Dr. Witlin laughed and said that "an overzealous co-worker of his had made that statement," which "wasn't true." However, he did not know where Reich's books might be obtained.

I didn't see any point in challenging Witlin's statement that the burning of Reich's books was not true. I was personally acquainted with the physician who reported loading the literature onto a city truck and seeing it destroyed.

Referring to me, Witlin suggested that I might be able to tell the audience where Reich's books might be obtained.

At that time, Marlboro Book Store on 8th Street was selling a few of Reich's works and I referred the audience there.

Toward the end of the evening, a member of the Humanist Society asked Dr. Witlin about the "legality" of other "qualified M.D.s" who were using the accumulator "when the accumulator has been outlawed."

The research director again rose and stated that the accumulator "had never been outlawed; anyone was free to build and use one, or even sell them."

Dr. Witlin said that this was correct, adding that the Department of Health "stood ready to act should anyone report a case of malpractice."

Now Dr. Witlin said that he had to leave, in spite of the fact that the chairman of the meeting informed him that there was still time for further questions. Dr. Witlin said that he was sorry, but he needed time to pack his contraptions and devices. "Perhaps some other time," he said. His final remark was that it had been "a very interesting evening."

11 The Atom Versus the Orgone

Those who instigated the attack upon Wilhelm Reich and Orgonomy did a thorough job. The loss of Reich's vital secrets to the United States is incalculable. The imprisonment and death of this scientific genius may yet prove catastrophic. One of the physicians who worked with Reich remarked: "The death of Wilhelm Reich spelled annihilation for humanity." This somber prophecy may prove accurate as our planet moves forward on a tightrope stretched precariously above a nuclear abyss.

Only a handful of scientists understand the functions of Orgone energy and its relationship to the atom. Since Reich's death, I have felt panic and deep despair as I watched the Frankenstein of nuclear energy burgeon, year by year, into a monstrous menace which can be triggered by a single hateful or resentful killer of life anywhere on the globe. Some believe it will require a nuclear war to bring man to his senses, to wipe this strife-torn earth clean (and barren) so that perhaps in another 10,000 years a newer and wiser form of life may again start all over.

Reich must have been tormented with deep agony over what hate-crazed men might do with his discovery of an energy more powerful than the atom's. He felt a tremendous responsibility for his work and for the living, too. The fact that he was repeatedly attacked by the leftist press, that his greatest efforts to bring his work to the attention of the American Government were continually blocked by pro-Communist sympathizers, plus the knowledge that the Soviet Union had acquired his publications, was utilizing his Orgone Energy Accumulator and had whisked his assistant and two motors behind the Iron Curtain—all of these facts must have contributed to his decision to publish and disseminate his findings.

The experimental work Reich undertook to evaluate the effects of the Orgone upon nuclear energy became the target for high-level espionage activities. Proof of this charge looms large throughout the entire court records and suppressed evidence. That area is the responsibility of the U.S. security departments which have complete access to such files and records. My single objective here is to focus attention upon the vital importance and magnitude of the work of Wilhelm Reich. When the *reality and scope* of Reich's discoveries gain sufficient public attention, it is my strong hope that America will act—before it is too late—in its own self-interest. As a consequence of this, the terrible injustice done to Reich will be acknowledged, and those who participated in the conspiracy to destroy him will have their day in court as well.

Now, let us see if we can comprehend Reich's Oranur Experiment, so that the "incredible" will become credible.

Nuclear energy and Orgone energy are *antithetical,* "mutually exclusive" forms of energy. Orgone energy is the primordial, mass-free energy out of which all matter arises. Nuclear energy, on the other hand, is energy derived from the *destruction* of matter. E equals MC^2 is a reversible equation. The "E" is Orgone energy.

For many years, while studying and elaborating on the functions of the Orgone, Reich deduced that atomic energy and the Orgone were antithetical, and he desired to set up an experiment to see what effects, if any, the Orgone, in high concentrations, would have upon atomic (or nuclear) energy. In particular, Reich was interested in the possibilities of uncovering any *immunizing* or *nullifying* effects which Orgone energy might have in this regard. He made very thorough and painstaking preparations for this experiment. In 1950 Reich was worried about the global crisis which might at any time engulf the United States in a nuclear war. This was the period of our epochal disaster at the hands of the Chinese Communists, who were throwing matches and supplying fuel to the holocaust in Korea.

In the middle of December 1950, Reich applied to the Atomic Energy Commission at Oak Ridge, Tennessee, for 20 millicuries of phosphorus-32 (P-32). In preparation for the arrival of this radioactive isotope, background counts of the normal radiation at various places around the Orgone Energy Laboratory and other buildings were recorded. The background counts in the Students' Laboratory, the site for the major portion of the experiment, were in an average range of 30 to 50 cpm (counts per minute).

So far as I can determine, this nuclear material ordered from the AEC either was delayed or never arrived at Orgonon. Reich thereupon ordered two milligrams of pure radium from a private laboratory. When the radium arrived, on January 5, 1951, Reich placed one milligram in its lead container in a garage some distance from the laboratory. This unit was to be the control. The second unit (one milligram of radium) was placed in a strong Orgone Energy Accumulator that was itself located in an Orgone Energy Room. After about five hours, the background counts began to increase steadily. Inside the Orgone Room the atmosphere had become intolerably charged.

Physicians and co-workers who participated in this experiment reported the same general symptoms: a saltiness on the tongue, feelings of pressure in the body, nausea, dizziness, and deep pressure in the epigastrium. Others experienced feelings of hot and cold flashes, or shivering. *The normally benign Orgone* energy appeared to have been converted into a "wild beast" with killer tendencies!

Work was thereupon suspended, and no one was allowed to enter the area where the nuclear material had been contained. Even with the removal of the radium, however, the building continued to remain "hot."

From January 6 to 12, the experiment was repeated, with the nuclear material being placed within the Orgone Room for one-hour intervals on each of those days. On the 12th, the one milligram of radium was placed in its Orgone "charger" and then in turn the charger was placed in the Orgone Room for only one-half hour. *Now all hell broke loose*!

Through the laboratory windows, bluish to purplish "clouds" could be seen. *Some 35 to 90 yards outside the building*, the participants began to suffer from severe nausea, dizziness to the point of fainting, and a loss of consciousness. *From the standpoint of classical physics, there is absolutely no reason why such a comparatively small amount of nuclear material would have such a deadly dangerous effect*! Nuclear energy can be shielded

—and *is* shielded—by metals or even concrete. Orgone energy, however, *cannot be shielded,* as it is able to penetrate everything, though at varying speeds.

A flood of incredible new data began to descend upon Reich and his research staff, all of whom became quite ill with "Oranur sickness." One physician nearly lost her life, Reich's young son suffered severely and had to be evacuated, and Reich himself was stricken.

The Oranur Experiment seemed at first to presage nothing but disaster. Blood tests (the "Reich Blood Test") were done on all participants and on laboratory mice. The blood of all co-workers appeared to suffer from intense overirradiation.

Despite the removal of the nuclear source, and the total dismantling of the Orgone Accumulators and Orgone Room, abnormally high background counts continued. Airing out the buildings and quarters repeatedly had a beneficial effect; however, when the windows were closed for only 15 minutes the cpm began to climb to 60 and 70! (Some of the participants were to suffer for months and even years after this experiment. As late as May 1951, Reich noted that the walls of the Orgone Room were still dangerously "hot," although the sheet-metal lining of this 18-by-18-foot room had long since been removed.)

Let us pause here for a brief recapitulation of what had occurred. The normally benign atmospheric Orgone energy had suddenly turned into a deadly killer energy after coming into contact with a relatively small source of nuclear material. The process can be compared as follows. I have a nice friendly cat—a benign, gentle feline who lives with me. (He is, as are we all, merely a sackful of Orgone energy.) Now, I take this friendly bundle of energy and goad him, say with needles. I prick him repeatedly, or I whip him with a strap. I beat him until he is driven to a frenzy of rage and killer hate! This *emotional* reaction is precisely what occurs *in the atmosphere* in the energetic realm of the atom's effect upon the cosmic Orgone energy. The reaction in the atmosphere, however, is not only locally energetic; it produces a *chain reaction, a slow infestation of highly charged, deadly Orgone energy that has become wild—in short, "Oranur"*!

This brief description of Reich's monumental Oranur Experiment is entirely inadequate to give anything but the barest outline of procedures. A torrent of new information regarding the pathology of "radiation sickness," of new insights into physical, emotional, and biological conditions, poured down upon Reich and his co-workers. Reich understood, moreover, after the initial onslaught of Oranur, that the Oranur Experiment *did in fact* demonstrate that Orgone energy had a *positive, immunizing effect* on nuclear radiation. Nuclear material kept for long periods within high concentrations of Orgone energy became "denaturized" and *benign*!

If administered in carefully moderated doses, Oranur would, Reich believed, prove invaluable *as a generally effective agent against nuclear radiation*! Furthermore, Oranur showed a special characteristic: *it attacked each person in his weakest point*, and was capable of bringing latent illnesses to the fore and of immunizing people against potential illnesses that might otherwise prove fatal.

I fully understand the difficulties that the reader must surmount in attempting to absorb and grasp this brief outline of the Oranur Experiment. And I am sharply aware of my own shortcomings in recounting this episode of pioneering research. For those who would pursue the matter further, I can only suggest referring to the Bibliography, knowing full well that the destruction of orgonomic literature makes it much more difficult to obtain the required technical information.

Now, let us look at a few implications of Reich's Oranur Experiment. We exist in an "ocean of energy"—Orgone energy. We also live in homes and work in offices that, knowingly or not, are in truth Orgone Energy Accumulators. Thus, many of us have been living and working in highly charged Orgone energy atmospheres for many years. Reich found, via this experiment, plus a multiude of subsequent observations, that nuclear testing —*regardless of where it occurs upon our planet*—has been contributing to *planetary* Oranur. Reich utilized only a small amount of nuclear material in a highly charged Orgone energy atmosphere. The result was Oranur. Exactly the same effect can be and is being produced by repeated nuclear testing, wherein *gigantic amounts* of nuclear radiation are released into the "normal" atmospheric Orgone energy. Remember, the Orgone energy is present *everywhere*—in the air above us, within our earth itself, and in our oceans, lakes, and streams. Thus, when nuclear physicists maintain that only a relatively small amount of radiation appears in our biosphere as a result of nuclear tests and nuclear reactors, they are correct; but what they don't realize is that such "minimal," "safe" amounts of nuclear material *are reacting upon an Orgone energy continuum* which has progressively become "activated" into Oranur.

Oranur sickness, which first struck down many of the participants in Reich's experiments, is now a worldwide threat to health. Failing to understand that every living being is in effect a *living* Orgone Energy Accumulator, our nuclear physicists and health physicists are totally unaware of the real hazards involved in "low-level radiation."

It is *not* the minute amounts of X-rays emanating from a TV set which are harmful; it is the "triggering effect" such radiation has upon the bioenergy of human beings which makes it dangerous. It is *not* the high voltage of a transformer that is dangerous per se; it is the Oranur effect it engenders in the living orgonotic system that is so deadly! It is *not* the low-level yields of our underground nuclear tests that are so dangerous; it is again the insidious Oranur effect that is harmful to human survival.*

Reich came to understand that many sources of *secondary forms* of energy, including radiation from *any* source—TV sets, ultraviolet light, high-voltage electricity, etc.—can produce the Oranur effect in human beings and in *all* living organisms.

Nevertheless, the Oranur Experiment does point the way toward the "Peaceful Use of Atomic Energy," if the scientific fraternity will strive to understand the functional realm of Orgone energy and *apply* Reich's Life-giving knowledge. Orgone energy can render nuclear energy *harmless*. Is this not a practical hope worth pursuing for the benefit of humanity and the cessation of the dreadful nuclear anxiety that hangs over our world?

Reich gave mankind a choice, an alternative to atomic suicide. God help us if we fail to grasp it!

---

*I suspect the recent epidemics of Infant Crib Deaths are due to the increase of Oranur and DOR.

12 No Place to Hide

The aftershocks of the Oranur Experiment continue to reverberate throughout the world. Reich demonstrated the relationship between the atom and Orgone energy. He continually kept the U.S. Government advised of his procedures. The Atomic Energy Commission received many reports from Reich's Maine laboratory, as did the Department of Health, Education and Welfare. Reich was fully aware of the international implications of the results of the Oranur Experiment. He offered on many occasions to cooperate fully with responsible representatives of the Federal Government, but no Federal offers or gestures of serious interest or cooperation came out of Washington. Instead, Reich was served with a court injunction which sought to discontinue his research and implied that he was somehow involved in quackery.

Millions of people throughout the world are becoming sick, and many are dying, from chronic Oranur and DOR sickness, as Oranur reactions continue to be produced everywhere on the globe via nuclear uses and from the noxious DOR clouds that are being poured into our living atmosphere from outer-space craft.

If, as I mentioned earlier, "radiation sickness" afflicted millions of people with some horrible disease, such as leprosy, the glaring aspects of such a disease could not be avoided. But the very nature of Oranur sickness is not that evident and certainly not that obvious to the ordinary layman and physician. Oranur sickness has *general symptoms* which are commonly referred to as "the flu" or "virus X": chronic weakness, wandering aches and pains, nausea, diarrhea, a need for water and oxygen, etc. In addition to this, Oranur (and DOR) hits each person in his weakest biophysical spot! And so old cases of hepatitis, rheumatism, gall-bladder trouble, migraine headaches, ulcers, etc., are reactivated and no one points the blame at Oranur! The real culprit, therefore, is generally never recognized, although more and more one hears people saying, "It feels like there's something wrong with the air!" or, "My, what funny-looking skies we are having!" Unequipped as society is with the facts regarding Reich and his discoveries, we face an immense task of education and orientation, with few capable of doing the job. I know of only *one* college, for example, that is presently conducting courses in the United States on the discoveries of Wilhelm Reich. That school is New York University at Washington Square, New York City.

Since the banning and burning of Reich's work and of the vast storehouse of orgonomic literature, I only know of *one* medical text that has been written on the subject of psychiatric orgone therapy: *Man in the Trap,* by Dr. Elsworth F. Baker. And there is only one scientific, technical journal being published in the United States (if not in the entire world) dealing with extensions and elaborations of Reich's basic discoveries: the *Journal of Orgonomy.*

Such a paucity of valiant endeavor in the face of the gigantic problems and dangers facing humanity on a planetary basis seems in contrast like the little lad who is attempting to hold back an ocean with one finger in the dike.

Many more scientists, physicians, and interested laymen must first become familiar with the gigantic problems involved. That is the purpose of this book.

The following occurrences, which I personally experienced, are given in an attempt to indicate the magnitude of the dangers inherent in a DOR and Oranur atmosphere. Most of this information appears in the May 1971 issue of the *Journal of Orgonomy,* in my article entitled "Personal Experience With Oranur."

Society tends more and more toward the mechanization of all existence. Machines are a great blessing, as anyone who has used both axe and chain saw to cut cordwood will attest. However, since the power supplied to run our machines is *secondary energy*—that is, energy derived from the primary Orgone energy ocean—and since, moreover, most sources of secondary energy irritate, and are antagonistic to, Orgone energy, mechanistic society rides precariously between the horns of a dilemma. On the one hand, it requires more and more machines to satisfy society's spiraling growth, while on the other hand, continued and increasing use of secondary sources of energy (e.g., high-voltage electricity, nuclear energy, gasoline) irritates the Life Energy into higher and higher states which are inimical and deadly to the living.

Over the years, I have come to realize how important my personal orgonotic (or bioenergetic) sense is to personal survival. Many persons have well-developed contact with their own emotions, though they tend to disbelieve or minimize their subjective feelings.

The continued increase of Oranur in our environment makes it especially important for the individual to learn to trust and to rely upon his subjective impressions, especially where Orgone energy functions are involved. Thus it occurred to me that a recounting of some of my own experiences with atmospheres that were contaminated by Oranur would be helpful to others who might tend to doubt the validity of their own subjective impressions.

The first incident I shall relate deals with an Oranur atmosphere engendered by high-voltage electricity. I was working as a sales engineer for a large company that manufactured laboratory equipment for hospitals and medical schools. My job was to demonstrate the use of this equipment, which included a microslide projector that used a mercury-vapor lamp as a light source. The machine required high-voltage electricity to light this lamp, which incidentally is rich in ultraviolet light, an additional irritant where Orgone energy is involved.

Invariably, whenever I worked in the vicinity of this projector I developed pounding headaches that circled my head at the ocular segment. Prolonged soaking baths and crying relieved the pressure. It did not take me long to connect the headaches with the projector. However, I could not avoid contact with this machine, as I was obliged to demonstrate its use several times a week.

While demonstrating the machine for physicians in poorly ventilated rooms, a few of the viewers fell sound asleep. Others acted lethargic. Instead of developing a tolerance for the projector (that is, for the Oranur atmosphere it "triggered"), I became increasingly sensitive to its influence. Eventually I left this job in search of other employment.

I obtained a job with an importer of wild animals whose offices were located in midtown Manhattan. I had discussed my inability to tolerate fluorescent lights with my employer, and he kindly agreed to replace those in my work area with incandescent fixtures. (Reich pointed out that fluorescent lights have an adverse affect on Orgone energy. Whatever adversely affects the Orgone *outside* one's body will of course have a similarly adverse effect upon the living organism.)

By this time, I was accustomed to my reactions to the irritating atmosphere of fluorescently lit spaces. They made me extremely uncomfortable, and the longer I remained in them the greater was my discomfort. My body grew cold, especially my hands and feet. I felt a constant need for oxygen, became very thirsty and often noticed mottling of my palms. It made little difference whether I worked directly under the fluorescent fixtures or was merely in their presence. The Oranur atmosphere was evident under such circumstance—although the closer I was to the fluorescent, fixtures, the greater the immediate effect. However, if I could work near an open window and was able to drink much water and lukewarm tea (as Reich recommended), and if I could bathe my hands and face and the back of my neck several times daily in cold water, I was then able to function to some degree. Nevertheless, my ability to work in such atmospheres was always sharply curtailed. I became increasingly irritable or dulled and found myself unable to do anything but rest at the end of each day.

But to return to my job with the Manhattan importer. It lasted exactly four days! In addition to the reactions described above, I found myself extremely anxious in this atmosphere, unable to think—as if I were being constantly harassed and "short-circuited." My reactions swung widely from deep contractions, with accompanying sweating, to severe, irrational anger at my employer. I could not understand my behavior. The situation did not appear to warrant such reactions; and being at a loss to explain it, I began to doubt my senses. My wife, however, pointed out the pallor of my skin and the black circles under my eyes. She suggested that I quit the job immediately. Our financial situation, however, was not good, and so I continued a second day, and then a third.

There was no denying my subjective sensations of severe anxiety, prolonged thirst, need for air, and inability to function. I knew I was in a severe Oranur atmosphere, but I was at a loss to explain the source. The fourth morning, I reported to work earlier than usual and examined the office carefully. Besides a few fluorescent fixtures and the usual metal file cabinets, I could locate no unusual triggering irritant to explain the immediate and overpowering discomfort I felt there. At lunch time I went into the street to air out. Returning to the office, I accidentally got off the elevator at the floor below my own. As I glanced around at the unfamiliar offices and workshops, the answer hit me: the entire floor beneath my office was occupied by a manufacturer of X-ray machines! I quit my job immediately.

The next incident was even more baffling. My wife and I had moved about 90 miles out of New York City and were managing a summer resort in the country. To supplement our income I took a job as a surveyor's apprentice with a professional engineer. I enjoyed the work immediately. My time was divided almost equally between field work in the woods and office work at the drafting table. My employer was a straightforward, considerate man, and we liked each other. I told him that I could not tolerate fluorescent lights, and he replaced the ones over my work area with incandescent fixtures. Nonetheless, in that office, *my subjective reactions of acute, immediate, and continuous discomfort were undeniably strong*! Each time I went into the woods and worked in the open air, I felt fine, well, and alive. But whenever I entered the office, I was smitten with deep anxiety, hot and cold flashes, and a constant feeling of harassment! By this time I could almost smell Oranur. But where?

Again, I was thoroughly familiar with the usual irritating atmosphere of fluorescent lights. Such an atmosphere was bad enough; however, the surveyor's office was ten if not twenty times more severe! I drank copious amounts of water and a thermos of coffee each morning. My face flamed with an inner heat, although my skin surface was cool. I felt stupid, thick, foggy, or bursting with anger or unaccountable giddiness. On any pretext I ran outside the office to get fresh air, or I bathed my hands and face in water. And whenever I was alone in the office, I searched for some source of my "antagonist," but to no avail. Was I going crazy?

I held onto my job for three months. Again it was a matter of financial need. Also, where could one get another opportunity to learn surveying at first hand? And where could one live and work in this society *without* fluorescent lights? I would be "tough"! I would "overcome" it! Each evening I looked forward to a prolonged lukewarm soaking tub, as Reich had recommended for Oranur victims.

But I found no immunity. Slowly as the days wore on, I felt myself becoming weaker. Undoubtedly the many hours spent surveying in the open air enabled me to hold out for those three months. But at length I felt I must again quit my job. It was a depressing decision, especially since I could not find the reason for my alarming subjective impressions of a highly charged Oranur atmosphere.

I was determined to search one more time. I drove to the Surveyors' Office on a Saturday morning, just to look around and say hello. The office itself was actually a suite of rooms in a large motel-hotel building. In the hotel office I asked the desk clerk if he knew who occupied the rooms immediately adjacent to the Surveyors' Office. (I don't know why this question had not occurred to me before.) The clerk replied that the adjoining office was vacant and had been for some time. I persisted; I asked for the identity of the last occupant of that office. The clerk referred to his files and then gave me his answer: for several years the office adjacent to mine had been occupied by a *radiological laboratory*! The other side of my own office wall was the location of the X-ray room, which had been lined completely, walls and ceiling, with lead! It was evident to me that the X-ray laboratory had unwittingly caused a severe Oranur reaction which persisted long after the source of radiation had been removed—exactly as had happened in Reich's Oranur Experiment!

I did my best to explain to my employer why I had to leave the job. That good man kindly advised me to place myself in the hands of a competent psychiatrist.

These experiences taught me very valuable lessons, especially since I was forced by circumstance to return to New York City. My next job found me working in a Manhattan office on Park Avenue. Fortunately, my employer saw fit to equip my office with incandescent lights at my request. The remainder of the entire office space on that floor was fluorescently lit. At last I had a private office with two windows. Nevertheless, the entire New York City atmosphere had become intolerable to me: the fluorescently lit metal subways, the metal and concrete buildings (acting as additional accumulating devices), the smog, automobile fumes, and high-voltage equipment everywhere made me feel as though I were constantly being "rayed."

Another inescapable irritant was television sets. I learned at first hand from an electronics engineer with a major company that TV sets give off low-voltage, or "soft," X-rays. My Orgone Energy Accumulator always became oppressive if placed as close as 30 feet to a TV set located in another room. On many occasions, while visiting someone who had a TV set in operation, the atmosphere was so unbearably charged that I would have to leave. My reactions were exactly similar to those felt while demonstrating the microslide projector, though not as intense.

Despite the fact that my private office contained incandescent lights, the general atmosphere was one of Oranur. One by one I collected twelve plants which gave me something green and alive to look at and which seemed to soften the charge. An electric fan placed in an open window and blowing air *into* the room brought relief. On very dry days, I turned on a vaporizer which can be purchased at any drugstore; the moisture aided also in softening the charged atmosphere.

At about the same time I obtained this job, my wife and I moved to an apartment in a new Manhattan building built over the 7th Avenue subway. The apartment was on the second floor. When we moved in, I left my Orgone Energy Accumulator, disassembled and protected by a canvas cover, on our open terrace. A few weeks later, feeling the need, I assembled the accumulator. After three days the atmosphere in the apartment was intolerably oppressive—undeniably Oranur. But why?

As I sat in my living room I heard a subway train rumble past, two stories or more below. On impulse, I got out my magnetic compass and waited for the next train to pass. As the train passed unseen two stories beneath me, *the compass needle was deflected 45 degrees!* I was certain that the high voltage of the subway system, plus the metal in the building and in the apartment, all interacted with the accumulator to cause the Oranur effect.

I disassembled the accumulator and re-placed it on the terrace. The Oranur atmosphere persisted, however, for several days. Fearing the accumulator had become permanently affected, I dismantled all the parts completely and had them burned. Even with the removal of the accumulator, the Oranur atmosphere continued for several weeks, despite frequent airing out of the apartment and washing down the walls and floors with cold water.

During the four-year period I lived and worked in New York City, there was no escaping the highly charged Oranur atmospheres. Finally, my wife and I decided to try our luck (for the second time) in the Alaskan wilderness, returning there in May 1966. That summer we took a trip into the Alaskan interior to revisit old friends and pick up several cartons of books being held for us. On the return trip I decided to rest on an open clearing off the Richardson Highway at about Mile 246. We had been driving all night and needed a few hours of rest to refresh ourselves.

I parked the car off the highway on a gravel clearing and spread out our sleeping bags on the ground. It was about 6:45 A.M., a mild, summer day. I remember admiring the unusually large, billowy clouds sailing dramatically by overhead. Directly behind us, about one-quarter mile away, was a single dome-shaped mountain.

We got into our sleeping bags and tried to sleep, but the sun seemed too hot and the mosquitoes too numerous. Suddenly I found myself unaccountably thirsty. My strong thirst was all the more peculiar because we had eaten breakfast only an hour or so before and I had consumed several cups of coffee and we had drained our canteen—thus we were temporarily out of water.

We decided to outwit both the mosquitoes and the sunlight by pitching our lightweight tent. In another few moments my wife and our German shepherd, Sergeant, were fast asleep inside. But sleep eluded me. I kept thinking about the water. My thirst was becoming an obsession, and I had to sit hard on my nerves to keep from flying into a panic. I told myself to be calm and reasonable—after all, we were only a few short miles from water. Why panic over a little thirst?

I lay on my sleeping bag and felt a continuous vibration shudder through my body. It felt like a truck engine was operating inside me. When I closed my eyes, all kinds of bright red and purplish blotches pulsated in my inner vision. My pulse began racing, I broke out in a cold sweat (despite the heat), and it was all I could do to keep myself from flying apart. The sight of my wife and dog asleep, however, reassured me somewhat, and this plus my tiredness overrode my survival instinct. I slept.

In a few minutes I was seated bolt upright, *filled with an undeniable panic!* My body felt parched, my lips were dry, my tongue was swollen. The "heat" seemed to have increased. I started to wake my wife, thought better of it, and returned to my sleeping bag. Again, the seeming gentleness and harmlessness of my surroundings, my wife's sleeping form, *and my unwillingness to believe my subjective sensations of severe distress* lulled me to sleep.

I do not know exactly how long I slept, but it was not longer than 30 minutes. I found myself sitting up, gasping for air. I felt as though I were dying of thirst, *actually drying up!* My heart pounded in my chest. The enginelike throb of my body had increased to such a degree that I was all but helpless. *I was experiencing the most severe death anxiety of my entire life!* Every bit of energy in my body was urging me to flee!

I shook my wife awake and told her we had to leave at once. She acted drugged and looked very pale. Sergeant appeared to have aged before my eyes—he appeared thin, emaciated. His head hung down, his eyes were nearly closed. I found it difficult to coordinate my movements and had to tell myself to put on first one shoe, then the other. Don't bother to pack anything, I told myself; just throw everything into the car and drive. Get away from here! Get fresh air, water! *Hurry!*

In a few moments we were on the highway again. The fresh air helped immediately. By the time we reached the next roadhouse, I had calmed down considerably, although the terrible thirst persisted. My wife and dog were sound asleep on the front seat. I filled our canteen, and all three of us drank freely. Then we washed ourselves and Sergeant with a soaked towel, and later we repeated the washing at a roadside stream. We felt much better.

It wasn't until several days later that I found the answer to this frightening event. A few friends came to visit us at our cabin. I asked them if they knew of any government projects where we had camped at Mile 246. Of course they did, they informed me. Everyone knew the "Donnelly Dome Area," the government's nuclear installation and atomic stockpile. The entire area is honeycombed with underground structures containing nuclear devices. Of course, my friends reassured me, they had seen the whole project on television and it was all "quite safe."

In March 1967 we moved to southeastern Idaho, where, as I mentioned earlier in this book, I found employment as city and military editor of the Idaho Falls *Post-Register*. Slowly we rebuilt our finances, bought a home and established ourselves in the community. We knew that the National Reactor Testing Station (NRTS) was some 40 miles west of our city. However, friends assured us that everything was, as usual, "quite safe" and there was no danger from radioactive pollution.

In the first section of this book I gave the factual information dealing with my criticism of the Atomic Energy Commission. I did not deal with my or my wife's *subjective* impressions, because it was first necessary to orient the reader into the realm of Reich's discoveries, and particularly Oranur. I will now proceed to our physical and emotional reactions to that highly charged Oranur atmosphere.

With all of the above experience behind me, I must admit frankly that I was still unwilling, or unable, to believe that the Oranur atmosphere *some 40 to 50 miles away from a nuclear reactor* could be so harmful! I have fled Oranur from the East Coast to the West Coast, and all the way up to the Arctic Circle. I had given up good jobs, excellent opportunities, and even careers. Now I had another excellent job in work that I enjoyed thoroughly (later, as mentioned in an earlier chapter, I became the director of my own agency). I was simply tired. And my wife was equally tired.

Thus, we lived for three years in that highly charged Oranur atmosphere, which finally became intolerable for us. As the months passed, we grew progressively weaker. Although we continued to use our Orgone Energy Accumulator blankets, they too became alarmingly weak. My wife, who never before had had a blemish on her face, broke out with a fiery-red skin rash. White scales continually formed on her eyelids. Our wonderful canine friend Sergeant developed a funguslike infection on his ears which five veterinarians could not diagnose. The disease began with the loss of hair on his ears and the appearance of a white "powder."

We watched, helpless and heartsick, as the tips of Sergeant's ears slowly dried and disintegrated. This despite 18 different kinds of medication used after Orgone energy "shooters" proved ineffective.

In order to *move*, I began taking cold baths (50 degrees Fahrenheit) each morning. On some days I would take a second cold plunge on my lunch hour. Inquiries among friends and acquaintances revealed that almost everyone was ill. Many complained of "chronic flu" symptoms. My wife became so weak that she often had to remain in bed for several days at a time. We slept with the bedroom windows wide open, winter and summer, and often used a window fan for additional fresh air.

Desperate, we began fleeing into mountain-and-tree country at every opportunity. (Southeastern Idaho is generally desert country.) The longer we remained away from the area, the stronger we felt, and my wife's skin rash would begin to subside. During our last six-month stay in this region, I began my campaign to find out why our area felt so deadly with Oranur. The result was my public battle with the AEC and the National Reactor Testing Station, which I detailed in the beginning of this book.

Thus, we were finally forced to move again. But we have not escaped the Oranur and DOR atmospheres even now. The sad fact is that there is no place on this planet that is not contaminated by Oranur and DOR to some degree. Civilized man has just about destroyed his basic planetary resources—the air, the water, the land. And he is, for the most part, totally unaware of what he is doing to the cosmic Orgone energy that is the foundation of all life on this planet.

We are now living on a 12-acre homesite in northern Idaho. It is probably one of the last remaining truly wilderness areas on our continent. Sergeant is still with us. He died four days after we moved into our new home, and we buried him at the edge of our clearing.

13 *Toward the Discovery of God*

> *He that speaketh of himself seeketh his own glory: but he that seeketh his glory that sent him, the same is true, and no unrighteousness is in him.*
> John 7:18

Man's recorded history gives worldwide evidence of an intuitive belief in good and evil, and in some supernatural power which is somehow responsible for either the good or the evil. Such a profound and persistent intuition, proclaimed by many religious prophets, must have its roots in reality. Reich often stated that everyone is "right" in some way, since what man experiences must be derived from some *basic reality,* inside or outside our bodies. It only remains for us to uncover the *ways* in which our prophets reflected this reality.

Reich was often deeply impressed by the "*meaningful*" behavior of the Orgone energy, especially in his Oranur Experiment, though he touched upon this aspect of Orgone functioning with extreme delicacy. It is apparent that he was as fearful of the *mystification* of his discoveries as he was of their abuse by mechanistic power-drunkards. Two of his published works, *Ether, God and Devil* and the companion volume, *Cosmic Superimposition,* deal with the basic functional law of our universe.

What follows now is my own summation of Reich's work which points to a factual approach toward an objective comprehension of the Creator.

Until the 20th century, knowledge of God was always considered beyond or outside the scope of human intelligence. From birth, man has been taught that although he is the child of God, still any *real,* factual knowledge of God is impossible to acquire. (God is "unknowable.") Such evasion of truth has inevitably developed into what can only be called "rationalized social insanity." Let us look at some of the symptoms:

Man, born of God, slays his brothers in bloody wars and riots fought in the name of their mutual Heavenly Father.

Man, "who cannot know God," nevertheless warps and crushes

the lives of billions of children whom he educates to fear God.

Every attempt to bring man to a knowledge of God and God's Natural Law has resulted in the persecution of the greatest benefactors of mankind.

Anyone who dared to live the life of Christ today would immediately be imprisoned or slain as insane, exactly as Christ was considered insane and was slain 2,000 years ago.

Whatever their religious differences, the world's religious leaders agree that man's happiness stems from knowing God and God's laws. Human misery is a consequence of ignorance or evasion of Natural Law. Until this century, any knowledge of exactly *what God is* was a matter of individual interpretation. A St. Jerome, for example, in his translation of the Bible, rendered the Latin word *caritas* into the word "charity," rather than give it the more exact meaning of "love."

At present, over much of the globe, mankind remains trapped in its religious ideologies (often disguised as political dogma) and supports its New Leaders who have rejected little of the ancient barbarisms. Today humanity suffers its new Political Religion (as Red and Black fascism) stripped of all subterfuge, decency, and humility. Such ideologies reveal nothing less than *the organization of malevolent power!*

Let us see if we can begin our factual search toward the discovery of God, of how our Creator functions in our universe and in our own bodies. Truthful, *factual* knowledge of God is indeed capable of "setting man free." Truth, however, is a two-edged sword. It can cut the bonds of the slave, but it cannot make him a free man. The slave himself must desire freedom, must be capable of containing freedom in his own body *without getting power-drunk,* and must assume the exacting demands and *responsibilities* of his freedom. *Nothing is more dangerous than the hate-enraged slave in power!*

The postulation of the existence of God has always included at least the following attributes or characteristics of some Creative Power to which the world's majority of religions subscribe:

A. God is the prime mover, or first cause, and all creation stems from this Creator;

B. God is Love and Life, and as the Life Principle, per se, maintains all that is alive upon our planet;

C. God is present everywhere throughout the universe, within and without the planets and His creations, including man;

D. God is expressed as the *Basic Natural Law* which governs our universe: expressed in the rotation and movements of the planets and celestial bodies, the law of gravity, the cycles of Nature (seasons, tides), and the growth and life-preservative behavior of living creatures (e.g., animal migrations, nursing the young, maternal love).

Let us for the time being limit ourselves to the above attributes or qualities of what mankind refers to as "God." Deliberately we omit *any moral considerations* of what man believes his Creator "wants" him to do. Such considerations have only resulted in splitting the human species into diverse and antagonistic religions, each of which claims that it is correct and all the rest are in error. If we are ever to be able to evaluate any code of morality, we must first strive to understand *what God is* before we can rationally attempt to comprehend *what God "wants us to do"*! How else can we bring order out of chaos?

Human history reveals a persistent hostility toward any person *who touches upon fundamental Natural Law*. Explanations such as "All great men have to suffer" are merely cheap excuses for terrible injustices. We have a right to expect, therefore, that the discovery of any factual, truthful evidence regarding Natural Law will be either evaded or suppressed.* Such truthful evidence should also reveal to us the answer to these crucial questions: Why has man thus far failed to know his God *in a practical, real sense?* Why has man persecuted, often to the point of death, all those who sought to bring him such knowledge?

Events of the past show clearly that anyone who comes close to "touching" God faces dire consequences. We must be prepared, therefore, to expect the following in the course of our search:

---

*See my correspondence with Idaho Senator Frank Church in chapter 14 for corroboration of this statement.

1. Factual, truthful knowledge of God will have a profound rooting in some area of human life which has been (and still is) considered, both individually and socially, to be worthy of deep contempt and murderous hate;

2. Factual knowledge of God will be intimately connected with man's profoundest terror and *irrational expressions;*

3. The pioneer discoverer of such knowledge, and his discovery, will be generally unknown; and where they are known they will be "officially" condemned—the discoverer himself branded as either "insane" or a "charlatan," treated worse than a common criminal;

4. Factual knowledge of God, expressed in Universal Natural Law, will be most violently opposed by all individuals and organizations whose existence depends upon ignorance or perversion of the Natural Law of God.

Thus, *evasion of the facts,* silent anxiety, or open hostility is to be expected from those who have always barred the pathway that leads toward truthful knowledge of our Creator.

Preliminary Approach

Where shall we begin our journey toward the discovery of God? What path shall we choose? Before us now is spread a map which traces the development of all human expression since the emergence of man. We see the multitude of threads tracing the growth of social and economic progress, the rise and fall of empires, the growth and decline of ideas, the phenomenal progress of mechanized technology. Men have spent their entire lives following but *one* of these tracings, which radiate outward and upward in a welter of divergent patterns, like branches on a tree, becoming the more complex and intricate the more removed they are from the single parent trunk.

Here we have religion with its variegated branches, ending in empty space. And here are mathematics, astronomy, medicine, biology, the arts—all branching from the parent trunk and all ending in empty space, fruitless in their search for God.

Everywhere we find allusions to a Creator, an *élan vital,* a

Universal Intelligence, a *Vis Naturalis*, an Ether, a Cosmic Consciousness. We find references to some "First Principle" in every branch of learning. Some refer to God as a "necessity," some fanatically. But nowhere do we find concrete, tangible, *objectively verifiable* evidence as to *what God actually is*! Our search along all established branches has proven fruitless. Thus we abandon the complicated branches and return to the parent trunk, and thence to the very roots of things—to man's origin in Life itself.

God Is Life

Although man is the earth's most adaptable and ingenious inhabitant, he may be a *carrier of creativity*, an *instrument* of creation, a shaper and manipulator. He may discover or destroy. God alone is the Creative Power, the "quickener" of Life. And so we take our first, tentative steps toward the discovery of God by asking: What exactly do we mean by *Life*? How do we know when something is alive? *What distinguishes the living from the dead?* Our answer is: We know that something is alive because *it moves*. Movement is an essential quality of the living. Dead matter is devoid of movement.

But there are many kinds of movement. Water in a brook moves; clouds move; the earth and the stars move; the mechanism of a clock moves. Obviously the movement of Life is quite different from mere mechanical motion or revolutions. Thus we arrive at our next question: What distinguishes *Life-movement* from mechanical motion?

It was Wilhelm Reich who discovered the specific formula that distinguishes the living. He expressed this specific Life-formula in this manner: Mechanical Tension> Charge> Discharge> Relaxation.

This formula is specific for the entire realm of Life. No matter what it is, every living thing, from ameba to elephant, expresses this basic four-beat pulsatory rhythm of Life.

The discovery of the Life-formula was a brilliantly luminating core of knowledge, one that is destined to play a primary role in our endless quest for God. Since this Life-formula is an expression of a basic Natural Law, by observing how it functions in Nature and in man we shall have gained our first firm foothold toward the climb to our objective.

The basic Life-formula (Mechanical tension> Charge > Discharge> Relaxation) can be observed microscopically in the tiniest of living creatures. When this fundamental expression of Living Life ceases, life itself ceases. The four-beat sequence of the Life-formula has many tempos. Its *pulsatory* expressions occur faster or slower, in individual cells, organs, and the various species of Life. In the living organism then, *Life is pulsation within a membrane*!

We now confront two important questions which arise from our study of the significance and pervasiveness of Reich's discovery of the Life-formula. If the Life-formula is *basic* to the entire realm of the living, *what causes* the pulsations expressed as tension, charge, discharge, and relaxation? In other words, we are asking for the *identity* of this moving, Life-creating, and Life-sustaining power. Can it be seen, felt, objectively verified? And if so, why has this force remained *unidentified* for so many thousands of years?

Let us pause briefly for some significant reflections. In the universe, the Divine Power (or what man refers to as "God") expresses itself in *creation*. No matter the differences existing among religious ideologies, those who maintain the existence of some Supreme Being agree that the Creative Power itself is responsible for human life and the perpetuation of all Life.

In man, God's creative process is most profoundly expressed in the genital embrace of male and female. This expressive "language of the living" is present within our universe in the *non-living* atmosphere. It is present also in the generative cells, and in the infant, child, and adolescent—*long before the sexual, procreative functions are developed!*

It is extremely significant that the Life-formula was discovered by Reich only after prolonged clinical investigation of the genital functions of the human organism which led to his discovery of the *function of the orgasm itself*. Until the work of Reich, man's knowledge of the creative function of the orgasm was either distorted or suppressed. The entire subject has a long history of evasion, suppression, and mystical or mechanistic misinterpretation. One finds here, too, the martyrdom of all pioneers for *rational* (that is, *healthy*) sexological enlightenment.

In brief, by treating his God-given genitality as the very "Devil," man has for millennia condemned himself to a virtual hell on earth. By refusing to touch this core of his life, man continues to close off the only avenue through which he lost Paradise and through which he must in fact return if he is ever to know his God as a living, objective—as well as subjective—reality.

It was Reich who discovered that *the orgasm formula is the Life-formula, per se,* operating within man as a fundamental expression of Natural Law.

Now we can begin to understand why the approach toward the discovery of God has been evaded by mankind all these years. Knowledge of God has been virtually outlawed from human society. Nevertheless, since man is truly "from God," every newborn child continues to bring with it into the world the very sap of Life. Thus mankind has always known of some quickening, enlivening influence in the cosmos, but it eluded humanity. Thus we have answered our second question as to why this power has *not* been identified and known for so many thousands of years. Now, before we proceed, one further point demands our attention.

Just as the Creative or Life Principle of our universe is considered an attribute of God, so too is the principle of Love. Man has held fast to his correct feeling that *God is Love and Life.* It remained for man to establish factually what is meant by "Love" and "Life."

God Is Love

The discovery that the basic, biological, *natural* (as distinguished from *secondary*, perverted, and pornographic) sexual expression of the human animal simultaneously expressed the specific Life-formula was received with typical human evasion—

silence, anxiety, or open hostility. During his lifetime of research, Reich's subsequent corroborative and scientifically verifiable work was based upon this single magnificent discovery which came almost unnoticed into a world of sick, floundering, and despairing human beings. Moreover, Reich's research demonstrated conclusively that *the majority of mankind suffers from a chronic incapacity to experience God's Love and Life in their own bodies!* Reich described the orgasm as "the unitary involuntary contraction and expansion of the total organism in the acme of the sexual act. This reflex, because of its *involuntary* character and the prevailing pleasure anxiety, is suppressed in most humans of today." (From *The Function of the Orgasm.*)

When Reich arrived at this description of the orgasm reflex, not only did he describe a profoundly Life-basic expression, but also he indicated the enormity and cause of mankind's self-perpetuating inferno.

The Prime Mover

We now return to our other question: *What is it* that motivates the human organism in its expression of the Love-Life formula? At first, Reich thought he was dealing with some type of bio-electrical energy acting in the body. Prolonged investigation, however, convinced him that the biological energy responsible for life was factually different from any energy hitherto discovered.

He named this biological energy "Orgone." He found that it was the Life Energy, per se, present in the atmosphere as cosmic, primordial energy.

Factual knowledge leading to man's discovery of God is established. The question awaits us in our immediate future: Will this generation finally succeed in beginning to know its God as a practical, tangible reality? Or will it destroy itself by continuing to support the advocates of hatred and death?

We are faced with a crucial choice which can no longer be evaded: *What will it be, atomic suicide or the Life-giving Orgone energy?*

14 *Whose Responsibility?*

Now, in this chapter, we have come full circle. Again the question rings out: "Who wants truth?" And the echoing question responds: "Why does mankind continue to evade new truths?" The answers cannot be found in history; they are, rather, problems of human emotional pathology, the pervasive illness of civilized man.

Each progressive step toward the betterment of man's lot, each inch forward in our long struggles toward a happy planetary existence, has been bitterly opposed. If my house catches fire and burns to the ground because I have no fire-fighting equipment handy, that is an act of ignorance. But if my house catches fire and I *refuse* to use the fire-fighting apparatus that is clearly in view, this is not mere ignorance *but criminal negligence!* To err out of ignorance is one thing; but to refuse to face facts (*reality*) or to deliberately evade, suppress or *destroy* the documentary evidence of such facts is simply criminal!

Nearly twenty years ago, when I first became interested in Reich's astounding discoveries, I sincerely believed that all one had to do was simply reveal such new truth to interested persons and they would be eagerly pursued. Reich's vital cancer work alone might save millions of persons from a terrible, painful death. Yet few pathologists even know of Reich's monumental work *The Cancer Biopathy*. I recall my excitement when I read this brilliant work. Being involved in the medical field, I showed the book to the director of a research laboratory whose very existence depended upon cancer patients. I can't forget his reaction. He opened the book, glanced at the title page and publisher.

"Who is this Wilhelm Reich?" he said disdainfully. "I never heard of him." His next remark was, "Hell! The book isn't published by a reputable publishing company!" Finally, after reading a brief passage, he slammed the book down on the table before us, exclaiming, "I haven't got time to read such crap!" That was this researcher's response to *years* of carefully documented experimentation of crucial importance in the field of cancer pathology.

It has often seemed to me that those who are supposedly most involved in certain areas of human problems are the very ones who obfuscate the problems or *block access to the solution of those problems.* That certainly sounds crazy, but the reception given to medical and scientific pioneers proves this to be true. Christ understood these machinations: "Woe unto you, lawyers! for ye have taken away the key of knowledge: ye entered not in yourselves, and them that were entering in ye hindered." (Luke 11:52.)

Wilhelm Reich and more than twenty other eminently qualified medical doctors and scientists published voluminous material on the new and vital science of Orgonomy. Whose responsibility is it to examine these important findings? Reich sent the results of his Oranur Experiment to many governmental agencies. Whose responsibility is it to study, duplicate and report these findings? Wilhelm Reich paid the supreme price in order to maintain his scientific integrity in the face of slander, ridicule, and deadly attacks. He more than discharged any responsibility he "owed" to humanity. Whose responsibility is it to further the difficult pathways Reich cut through our tangled jungles of sickness and social miseries?

For many years I have attempted to interest various state and Federal Government officials in the discoveries of Reich. My latest attempt was concluded on February 20, 1971, with a letter from U.S. Senator Frank Church. Let us examine this correspondence, which actually began in 1970 when I was criticizing waste-disposal practices of the National Reactor Testing Station near Idaho Falls.

The latest exchange between Senator Church and myself began with a letter I wrote him as follows:

December 26, 1970

Senator Frank Church
United States Senate
Washington, D.C. 20510
Dear Frank:

Thank you so much for your letter of December 10th. My wife and I wish you and your family the best for the Holiday Season and the coming New Year.

You may recall my personal battle with the A.E.C. when I lived in Idaho Falls, and the material I sent you detailing the reasons why I believe the atomic energy industry constitutes a major hazard to the people of southeastern Idaho.

Having moved here to northern Idaho, we find, as I'd hoped, that our health is returning. . . .

I am fully aware of the general need for developing atomic energy—for the maintenance of America's role in the world and to supply the growing power (electrical) needs of our nation, despite the attendant hazards. There is, however, *an alternative*—a source of power infinitely greater, which is NOT life-inimical and which could usher in a new Dawn of Hope for mankind.

I refer you to a recent book, *Psychic Discoveries Behind the Iron Curtain,* by Ostrander and Schroeder, published this year by Prentice-Hall, Englewood Cliffs, N. J. The term "psychic" is an unfortunate one, as the USSR knows full well about the work of the late Dr. Wilhelm Reich,* who died in a U.S. Federal penitentiary in 1957, and whose discovery of the cosmic Life Energy (the antithesis of nuclear energy) holds the key to world power and survival.

All I will say at this point is that there is a small group of highly qualified and dedicated scientists who are heroically working to advance the knowledge and use of this energy. To date, the work points positively to the solution of our most urgent problems: a reversal of atmospheric pollution, desert and drought development, and an answer to a practical substitute for atomic energy.

I could send you a carton of published verification, but the matter would be better explained by a personal, friendly exchange, away from the pressures of your office. So, if you get to this neck of the woods, I would be happy to discuss this all-important issue with you and explore possible U.S. approaches.

Thank you for your deep concern for our state and nation.

Our best to you and your family for the New Year.

JEROME EDEN

On December 31, 1970, the following telegram was received from Senator Church:

**Psychic Discoveries Behind the Iron Curtain* is a very exciting book and definitely shows that the Russians are working with orgone energy and know it. One of my patients attended a lecture given by one of the authors. She [the author] said they [the Russians] have everything he [Reich] has ever written. (From a personal letter received from Dr. Elsworth F. Baker.)

REURLET: SHARE YOUR CONCERN RE ATOMIC WASTES. YOUR KIND WORDS APPRECIATED.

FRANK CHURCH

That same evening I sent the following letter to Senator Church:

December 31, 1970

Senator Frank Church
United States Senate
Washington, D.C 20510

Dear Frank:

Thank you for your telegram this evening. Your response to my letter of December 26 was, however, a disappointment to me. So I've decided to try again to interest you—and through you our Government—in a matter of the deepest import to our nation, if not to our entire planet.

May I please remind you of my background. . . .

I know you to be a courageous man in dealing with the facts regarding radioactive hazards. I have read your remarks in the *Congressional Record.*

Our nation (and our planet) is confronted with a terrible dilemma: we need atomic power for economic growth, and, simultaneously, this same atomic power threatens us with extinction.

Years ago, when our country faced a military crisis, Albert Einstein wrote a letter to President Roosevelt sketching the discovery of a most potent force: atomic power. That single letter changed the course of history.

In all humility,* I am suggesting that this letter—in this day and time—can alter human destiny in like regard.

Wilhelm Reich, M.D., a naturalized American citizen, worked for many years, quietly and without fanfare, and discovered the cosmic Life-Energy (which he termed "Orgone" energy—after "organic"). In quietly published books and journals, many highly placed physicians and scientists corroborated his work. I have all of that published material—including photostats of letters between Einstein and Reich. Orgone energy is the antithesis of nuclear energy; and the Orgone is by far the more powerful.

To be brief, Dr. Reich was brought up on fraudulent charges by certain enemies of our nation who used a U.S. agency and U.S. courts to do two things: destroy further research in Orgonomy and discredit Reich as "insane." Reich fought valiantly,

*As subsequent events unfolded, I should have added "and naivete"!

but nevertheless went to a Federal penitentiary, where he died November 3, 1957, a week before he was to be released.

One of Reich's assistants disappeared behind the Iron Curtain with two motors that Reich had invented, which utilized the Orgone as propulsive power.* It is known, too, that certain devices invented by Reich are also being used in the Soviet Union. The recent book, which I mentioned in my letter of Dec. 26, indicates that the USSR is utilizing Reich's discovery.

Reich's discoveries offer, as I said, a new Dawn of Hope for mankind, and especially our nation, providing the Government is willing to carry the work forward. It means a new, infinitely potent source of propulsive power. It means a practical means of national and global *weather control,* already proven feasible and technically reliable. It means a reversal of drought and desert development; it means an affirmative answer to a substitute for atomic power, far safer and Life-positive.

It means, furthermore, entire new industries and instrumentation utilizing this cosmic Life Energy, and a practical response to our encroaching pollution problems.

If this incredible source of power should fall into the hands of our enemies (and it looks like they have already beaten us to the starting gun here) we might as well hand over the keys to the White House!

Now, it's New Year's Eve, Frank—and I am sitting here in a little dinky town (that's probably not on anyone's map) . . . one man with a typewriter. And I say to myself, Frank Church cannot believe me. Why should he believe me? He will think I'm just some kind of kook.

Let us therefore pretend the situation is reversed: *I* am the United States Senator, and you are the ordinary man in northern Idaho. *You* have all the Reich's published records; you've studied the court trial, you have parts of the court records; you've duplicated many of Reich's original experiments. . . . WHAT WOULD YOU DO?

My wife—God bless her—tells me it's hopeless.

Maybe it is, Frank. This issue, should it ever come before congressional scrutiny, will make and break many a leader. That I promise you.

In my book, Wilhelm Reich was one of the greatest men to walk this earth. His death was a national disgrace. The burning and banning of his scientific works in August 1956 was horrifying. And the theft of his discoveries by a foreign power will be catastrophic, "unless the right man does the right thing."

---

*Actually Reich did not invent the *motors,* but discovered the way to run them on Orgone energy.

If you are interested, look up WILHELM REICH, THE WILHELM REICH FOUNDATION AND MICHAEL SILVERT, petitioners, versus United States of America, respondent: Petition for a Writ of Certiorari to the United States Court of Appeals for the First Circuit . . . Supreme Court of the United States, October Term, 1956, No. 688 (Washington, Jan. 10, 1957).

I've put the torch in your hand, Frank.

May God give you the wisdom to know the right and act on it.

My best wishes to you and your family for the New Year.

Most sincerely yours,
JEROME EDEN

About one month later, I received the following note from Senator Church:

January 21, 1971

Dear Jerry:

Thank you for your most recent letter regarding energy sources other than those receiving current recognition in our country.

I appreciate your time in passing this information on to me. I have taken the liberty of forwarding your letter without use of your name to the National Academy of Science here in Washington for a report on your comments. When I receive that report I will be back in touch.

With best wishes,

Sincerely,
FRANK CHURCH

On February 20, 1971, I received the following letter and enclosure from Senator Church:

February 16, 1971

Dear Jerry:

Enclosed is the reply that I have received from the National Academy of Sciences in response to my inquiry on *your behalf* concerning "Orgone energy." I hope this reply is helpful to you.

With best wishes,

Sincerely,
FRANK CHURCH

For the reader's benefit, I underscored the words "your behalf" in Senator Church's note above. I was not asking Senator

Church to make inquiries for *me*. I am asking Senator Church, as an elected public official, to investigate this crucial matter for *himself* and for the welfare of his state and nation which he serves.

In anticipation of the reply from the National Academy of Sciences, I drafted the following letter to Senator Church, which, however, I did not send to him.

January 25, 1971

Dear Frank:

Your letter (of Jan. 21st) re the vital energetic discoveries of Dr. Wilhelm Reich was a blockbuster of naivete. [I might add that my own letter to Senator Church was an equivalent "naive blockbuster!"]

You are "passing this information on to the National Academy of Science for a report on [my] comments." I don't mean to insult you, Frank—please understand me. But just consider the following for a moment:

What kind of "official report" would you have gotten in 1902 from the NATIONAL ACADEMY OF HOT AIR BALLOONS regarding the discoveries of the Wright brothers?

What kind of "official report" would you have gotten in 1600 from the ITALIAN FRATERNITY OF SCIENCE regarding the discoveries of Galileo?

What kind of "official report" would you have gotten in 1889 from the AMERICAN CANDLEMAKERS INSTITUTE regarding the incandescent light bulb of Thomas A. Edison?

Nobody—but NOBODY—regardless of degrees and academic position is qualified to pass judgment on NEW FIELDS OF RESEARCH unless he has personally *done the work* in the field under scrutiny!

Now—you want to know what kind of report you will receive? (If not, I'll happily chew it up and swallow it!) This is what you'll get:

Your inquiry has been referred to our Investigative Committee which looked into the background and history of (Galileo, the Wright brothers, Edison, and now Dr. Wilhelm Reich). A careful analysis of his research and claims leads us to the belief that the gentleman in question is obviously deluded, has no basis in scientific truth for his claims, is most likely operating for personal notoriety and/or financial gain.

An investigation undertaken in (1600, 1889, 1902, etc.) by the Scientific Purity Federation indicates that the charlatan(s) in question was milking the public, preying on their fears, promoting worthless devices, and making fraudulent claims. Therefore, for the protection of the public, we find that the Purity Federation seized their publications, destroyed their experiments, and, wherever practicable, made the recalcitrant frauds recant—either publicly or privately. In some instances, Federal incarceration was undertaken.

We trust the above sufficiently answers your questions. We feel you would be well advised to discontinue further investigation into this obviously fraudulent individual and his questionable activities We appreciate this opportunity to be of service to you, Senator, and through you to your constituency. If you have any further queries regarding (Galileo, the Wright brothers, Edison or Dr. Wilhelm Reich) we shall be happy to reply to them with our usual thoroughness.

Most respectfully yours,
Arnold Benedict Hatchet, DhP, Rx. DDT
Director,
Closed Curtain Investigations

The actual response from the National Academy of Sciences was somewhat different. The results of the response were, for all practical purposes, the same:

January 28, 1971

The Honorable Frank Church
United States Senate
Washington, D.C. 20510
Dear Senator Church:

To the best of my knowledge, "Orgone energy" *has never been demonstrated in a way comprehensible or measurable by a physical scientist*. We therefore question such statements as "proven feasible and technically reliable." [Italics added.]

If such evidence exists in the form of specific data, it would be helpful in evaluating the claims of your constituent. *Lacking such evidence, we are not able to comment constructively.* [Italics added.]

Sincerely yours,
John S. Coleman
Executive Officer

This is a superb tactic in the gamesmanship of orthodox science! It is interesting, is it not, that the Executive Officer of the National Academy of Sciences does not have access to the voluminous literature on Orgonomy that was sent to the United States Library of Congress, the Department of Health, Education, and Welfare, the Atomic Energy Commission, the Federal Bureau of Investigation, the United States Air Force, or deposited with the American courts?

Thus, *without perusing the voluminous published literature on Orgonomy,* and, apparently without making the effort to find such literature (he *could* have asked the *Soviet Embassy*!), Executive Officer John S. Coleman of the National Academy of Sciences nevertheless goes on record as stating that Orgone energy *"has never been demonstrated in a way comprehensible or measurable by a physical scientist"!* Wasn't Dr. Albert Einstein a physical scientist? The medical doctors, mathematicians, biologists, and chemists who published corroborating papers attesting to the scientific validity of Orgone energy—were they not, and are they not, physical scientists?

And it must be pointed out that although Executive Officer John S. Coleman—owing to "the lack" of physical, supportable evidence—is "not able *to comment constructively,*" nothing apparently prevents Executive Officer John S. Coleman from *commenting destructively!*

"If such evidence exists," says Mr. Coleman. *If it exists!*

As published in "Petitioner's Reply Brief," No. 688: Supreme Court of the United States, October Term, 1956: Wilhelm Reich, et al., versus United States of America:

The following scientific works went up in flames between June and August of 1956, on order of agents of the food, drug and cosmetic agency. *Are you listening, Executive Officer John S. Coleman?*

1. The Function of the Orgasm, 1942 (banned)
2. The Cancer Biopathy, 1948 (banned)
3. The Orgone Energy Vesicles (Bions), 1942
4. Experiment XX, 1945
5. The Visualization of the Atmospheric Orgone, 1942
6. The Mechanism of the Concentration of the Atmospheric Orgone, 1943

7. Thermical Orgonometry, 1943-44
8. Electroscopical Orgonometry, 1943-44
9. Quantitative Determination of the Orgone, 1944
10. The Carcinomatous Shrinking Biopathy, 1942-43
11. Orgone-Biophysical Blood Tests, 1942-43
12. Anorgonia in the Carcinomatous Shrinking Biopathy, 1945
13. Orgonotic Pulsation, 1944
14. A Motor Force in Orgone Energy, 1949
15. Further Characteristics of Vacor Lumination, 1949
16. Public Responsibility in the Early Diagnosis of Cancer, 1949
17. Cosmic Orgone Energy and "Ether," 1949
18. Ether, God and Devil, 1949
19. Orgonomic and Chemical Cancer Research: A Brief Comparison, 1950
20. Orgonomic Literature Ordered From Russia, 1950
21. On Scientific "Control," 1950
22. Orgonomic Equations: 1. General Form, 1950
23. Meteorological Functions in Orgone-Charged Vacuum Tubes, 1950
24. The Orgonomic Anti-Nuclear Radiation Project (ORANUR), 1950
25. "Cancer Cells" in Experiment XX, 1951
26. The Anti-Nuclear Radiation Effect of Cosmic Orgone Energy, 1951
27. Complete Orgonometric Equations, 1951
28. The Storm of November 25th and 26th, 1950, 1951
29. The Leukemia Problem: I. Approach, 1951
30. The Orgone Energy Accumulator, Its Scientific and Medical Use, 1951
31. Armoring in a Newborn Infant, 1951
32. "Dowsing" as an Object of Orgonomic Research (1946), 1951
33. Three Experiments (1939), 1951
34. Wilhelm Reich on the Road to Biogenesis (1935-1939), 1951
35. The Oranur Experiment: First Report (1947-1951), 1951

36. My Experience With the Orgone Accumulator, 1949
37. The Treatment of a Mediastinal Malignancy with the Orgone Accumulator, 1949
38. Orgone Therapy of an Early Breast Cancer, 1950
39. Limited Surgery in Orgonomic Cancer Therapy, 1951
40. Again the Common Cold, 1947
41. Emotional Plague Versus Orgone Biophysics, 1948
42. Orgonomic Treatment of Ichthyosis, 1951
43. Orgone Therapy—a lecture delivered before the Medical Students Union, Oslo, Norway, 1950
44. Orgone Therapy in Rheumatic Fever, 1950
45. The Treatment of a Hypertensive Biopathy with the Orgone Energy Accumulator, 1951
46. Wilhelm Reich's Concept, "Cancer Biopathy," 1950
47. The Second Law of Thermodynamics and the Orgone Accumulator, 1949
48. Mathematical Questions Without Answers, 1951
49. Life and Matter. A Series of Three Experiments, 1938
50. Physical Orgone Therapy of Diabetes, 1951
51. From the History of Science, 1951

If painstakingly elaborated and published scientific findings over a period of 30 years could not convince this administration, or will not be able to convince any other social administration of the true nature of the discovery of the Life Energy, no litigation in any court anywhere will ever help to do so.—*Wilhelm Reich, M.D.* (From Reich's "Response to the FDA Injunction.")

15 *The Final Fight for Life*

Today the human species faces the greatest crisis in its entire fearful history. This dangerous period is all the more threatening because of the subtle manifestations of the threat of which so very few are aware.

Whether man and *all* planetary life will survive the next few decades (if, indeed, we have even *that* much time) will depend entirely upon our ability to face our problems squarely, utilize our best brains in assessing the situation, and adapt our technological and emotional behavior in conformance with the new facts and insights given us by Wilhelm Reich and his discovery of the Orgone energy.

It is well known that mankind cannot survive a nuclear holocaust. Orgone energy is far more powerful than the atom. Its beneficence, however, is matched only by its equal power to be converted into a deadly killer, as Reich's Oranur Experiment clearly demonstrated.

Any classical physicist or scientist who doubts the validity of the Oranur Experiment can easily be convinced of its reality by sticking his head into an Orgone Energy Accumulator that has become contaminated by Oranur. The resultant personal reaction should knock any further doubts out of his head! The procedures for such an experimental setup are given in Reich's Oranur Experiment.

Let us therefore proceed with caution and wisdom. But let us by all means *proceed immediately!*

What Can Be Done?

The following proposals are offered as a working outline only of steps which can be implemented in reversing our present trend toward atomic suicide:

1. The U.S. Atomic Energy Commission should immediately be transformed into the U.S. Orgone Energy Commission (U.S.OR.C.), to be headed by physicians and scientists who have *established working credentials* in this vast new field.

2. The U.S.OR.C. should be charged with the responsibility of the following:

   A. To study present atomic energy plants and installations and make recommendations as to their present or potential hazards in the light of Oranur;
   B. Establish immediate educational programs to orient public and private industry in Orgone energy functions —specifically those industries utilizing nuclear radiation, X-rays, high-voltage electricity, etc., which are now, as a consequence, producing the Oranur effect;
   C. Establish educational programs to orient medical and health authorities in Orgonomy, at the medical school, university, and hospital levels; national crash programs on orientation in Orgonomy would also be instituted, once more and more scientists, engineers, physicians, and educators were themselves grounded in this new field.

3. The U.S.OR.C. would establish a Cosmic Orgone Engineering Department (CORE), with responsibility for the following:

   A. Conducting research into Reich's weather-control work;
   B. Investigating the source of present planetary DOR infestations, which continue to threaten national and planetary drought and desert development; this would include the development of national "cloud-busting" stations in a unified effort at control of forest fires, drought, and smog removal;
   C. Investigating the problems posed by outer-space craft in our atmosphere.

4. The U.S.OR.C. would establish a Department of Medical Orgonomy, headed by trained medical orgonomists who would develop a national program of research and education, utilizing Reich's basic discoveries of Natural Law, the Orgone Energy Accumulator, the Medical DOR Buster, the Cancer Biopathy, Oranur, Reich's Blood Test, etc.

The above proposals are merely a rough working outline of some of the most urgent organizational work which must be accomplished first. Many problems are to be expected, and many will arise. Nevertheless, we cannot get bogged down at the outset in objections, questions, and procedural details. As Reich pointed out, had Columbus stopped to count every single nail required to build his three ships, he would never have put to sea!

A final word should be spoken relative to the court case against Reich and Orgonomy. The injustice done to truth and fact, and the death of Wilhelm Reich, were national and human disasters. Millions will needlessly suffer and die because this vital new field of research was crushed to earth and a magnificent scientist was destroyed.

An immediate *full and open* investigation into the Reich case should be instituted by the Congress. Those found guilty of breaching a public trust should be brought to trial and punished, in accordance with the laws of boomerang justice laid down in Reich's reply brief. But this alone is not enough to secure *future* justice and safety. A close scrutiny of the Reich case, the court records, the suppressed records, etc., will indicate what measures must be taken to safeguard democratic human and scientific freedoms for our present and future generations.

Responsibility of the Public

Like it or not, in any society *the people*, who constitute the working power and the paying power, get what they desire. They get it either because they openly approve what is going on around them or because their *silence* is always construed as *approval*! It takes only a few moments for any individual to write a letter to his congressman, senator, governor, etcetera. Your *one letter*, multiplied by the thousands or millions, cannot be ignored by your representatives. Only when you, as an individual, demand truth and justice, will our nation obtain truth and justice.

In the light of this impending atomic crisis that threatens all of us with extinction, let us put aside our partisan resentments and differences and think of the children—*all children of our nation and our earth*, black, white, red, yellow. All children function on the *blue* Orgone energy—our Creator makes no distinction as to color, class, or education. Let us focus our attention, therefore, on the Almighty Creator, and in attempting to comprehend the Functional Laws of Life, we shall in turn receive the gift of Truth which can truly "set men free."

Epilogue

An Open Letter to Youth:

In these last decades of the 20th century, a new world has been emerging. One sees everywhere the evidence of this struggle of birth—the misery and frenzy of the old world dying amid pangs of suffering, destruction, and death.

Youth is the center of whatever new world awaits us beyond the dark horizon. On the credit side, youth has the fresh strength of new life, the ability to desire necessary change, and the courage to work toward such change. On the debit side, youth lacks the patience, wisdom, and humility that come only from practical experience through love, work, and knowledge.

Throughout the bloody history of Earth, youth has been sold into private and public servitude, sacrificed as pawns in international chess games, flattened into mechanistic gears to fit the patterns of politics, castrated into conforming to rigid, self-serving segments of society, conscripted into the abattoirs of military machines, or imprisoned in madhouses or penal institutions, abandoned and shattered.

Still, I say, youth is the Hope of the Earth, the one last Light of the World. And therefore these closing remarks are directed primarily to the young people of our beleaguered planet.

Nothing rational, creative, healthy, or socially useful—*nothing, I say—can be accomplished by irrational, destructive, sick, or contemptuous means*! You cannot grow potatoes by planting dynamite. No matter how many tons of dynamite you plant and detonate, you will not grow one solitary spud thereby! This is such an obvious truth that it appears at first glance to be an absurdity. Who would counsel you to create a garden, a home, or a new world by planting dynamite or sowing Molotov cocktails! Yet this is exactly what the Enemies of Mankind would have you do.

Such killers have no more concern for your welfare than rattlesnakes have for rabbits. The name of the game is Power; and yours is the living energy that feeds it. The snakes will poison you (with propaganda or pills), paralyze your creativity, suck dry your life, then abandon you for new, fresh rabbits when you are no longer "socially useful."

Before it is too late, learn to recognize such Killers of Life who conspire to ensnare, enslave, and destroy your young, power-wielding strength. Learn to read the *emotional expressions* of a humanoid viper, a jackal, an unctuous fox, or a filthy vulture.

Before you blindly follow this or that "New Leader," ask simple, direct questions; and *demand* simple, straightforward answers! What practical, constructive, creative *work* has the glib New Leader performed? Has he ever built a home with his own two hands, worked a farm, raised healthy, happy children, or even healthy chickens? Or is he merely another empty, political windbag—a pestilential blowhard who does nothing except condemn, ridicule and destroy?

One does not enlist the services of an architect on the basis of his contempt, his ability to denigrate or to plant bombs in washrooms! One asks of an architect, *especially a social architect*: "Show me your concrete, practical plans for what you desire to *build*! Let me examine a structure you have already erected!"

Nothing is more deadly to your life than the hate-enraged Little Man who claws his way to power on the backs of the bewildered and helpless masses. Wilhelm Reich taught us this bitter truth; and he gave you many more vital tools to use in this final Battle for Life.

Yes, you have grievances, many painful, legitimate grievances. I have been in contact with your misery for a very long time. I was for several years a public-school teacher; and my book *Suffer the Children* attests to my deep knowledge of your legitimate resentments. In 1959 I was fired from my profession as an educator for attempting to come to grips with your desperate struggle for life and love. That was long before protesting became the vogue. Nobody marched down Main Street or picketed to protest my dismissal—just as nobody organized protest marches against the shocking arrest, conviction, imprisonment, and death of Wilhelm Reich.

You are not the only ones who have been slapped in the face, spat upon or wronged! *There is not an adult alive today who has not undergone the Crucifixion of Life*! Urinating on public monuments and burning the national flag, however, do not alleviate the Crucifixion of Life—they only implant more murderous seeds!

Some 200 years ago, another great man, Dr. Franz Anton Mesmer, wrote: "Truth is a straight line drawn between error." Our present social violent deviations are leading us inevitably to the far left or to the far right of totalitarianism. The dedicated, straight path of creative life is often slow, painstaking, requiring much discipline and patience.

What will your future be like if at present you are unable to distinguish between good and evil, healthy and sick, God and Devil, love and perversion, vitally necessary work and the parasitic prostitution of youth?

How can you learn to make such crucially important distinctions? For a start, I sincerely recommend the long, love-given labor of Wilhelm Reich.

May you have the wisdom and courage of God—which is the rational Life and Love in your bodies—to be guided by the true stars of rational progress: Love, Work, and Knowledge.

The flame of Life is truly in your hands. You are not alone in the long struggle for a better existence. I fervently pray you will use your God-given Life Energy to luminate and not destroy the last vestige of hope that still remains on the crippled planet we must continue to share.

Most sincerely,

Jerome Eden

Appendices

Appendix A

February 25, 1954

The Honorable Judge Clifford
Federal Court House
Portland, Maine

Dear Judge Clifford:

I am taking the liberty of transmitting to you my "Response" to the complaint filed by the Food and Drug Administration regarding the Orgone Energy Accumulator. My "Response" summarizes my standpoint as a natural scientist who deals with matters of basic natural law. It is not in my hands to judge the legal aspects of the matter.

My factual position in the case as well as in the world of science of today does not permit me to enter the case against the Food and Drug Administration, since such action would, in my mind, imply admission of the authority of this special branch of the government to pass judgment on primordial, pre-atomic cosmic orgone energy.

I, therefore, rest the case in full confidence in your hands.

Sincerely yours,

/s/ WILHELM REICH, M.D.

RESPONSE

Regarding the Request of the Food and Drug Administration (FDA) to Enjoin the Natural Scientific Activities of Wilhelm Reich, M.D.

In order to clarify the *factual* as well as the *legal* situation concerning the complaint, we must, from the very beginning, distinguish concrete *facts* from *legal procedure* to do justice to the facts.

Technically, legally the US Government has filed suit against the natural scientific work of Wilhelm Reich.

Factually, the FDA is NOT "The US Government." It is merely one of its administrative agencies dealing with Foods, Drugs and Cosmetics. It is not empowered to deal with *Basic Natural Law*.

ORGONOMY is a branch of BASIC NATURAL SCIENCE. Its central object of research is elucidation of the Basic Natural Law.

Now, in order to bring into line the legal procedure with the above-mentioned facts, the following is submitted:

The common law structure of the United States rests originally on Natural Law. This Natural Law has heretofore been interpreted in various ways of thinking, metaphysically, religiously, mechanistically. It has never concretely and scientifically been subjected to natural scientific inquiry based upon a discovery which encompasses the very roots of existence.

The concept of Natural Law as the foundation of a secure way of life, must firmly rest upon the practical concrete functions of Life itself. In consequence, a correct life-positive interpretation of Natural Law, the basis of common law, depends on the *factual* elucidation of what Life actually is, how it works, what are its basic functional manifestations. From this basic premise derive the claims of natural scientists to a free, unmolested, unimpeded, natural scientific activity in general and in the exploration of the Life Energy in particular.

The complaint of the FDA is factually intimately interconnected with a basic social issue which, at present, is reverberating in the lives of all of us here and abroad.

Abraham Lincoln once said: "What I do say is that no man is good enough to govern another man without that other's consent. I say this is a leading principle, the sheet anchor of American republicanism."

At this point, I could easily declare, "I refuse to be governed in my basic natural research activities by the Food and Drug Administration." But exactly here, in this constitutional right of mine, the basic confusion in the interpretation of Natural and Common Law becomes apparent.

There are conspirators around whose aim it is to destroy human happiness and self-government. Is now the right of the conspirator to ravage humanity the same as my right to free, unimpeded inquiry?

It obviously is NOT THE SAME THING. I shall not try to answer this basic dilemma of American society at the present. I shall only open an approach to this legal and factual dilemma. It has a lot to do with the position of the complainant, trying to enjoin the experimental and theoretical functions of Life in its emotional, educational, social, economic, intellectual and medical implications.

According to natural, and in consequence, American Common Law, no one, no matter who he is, has the power or legal right to enjoin:

The study and observation of natural phenomena, including Life within and without man;

The stir to mate in all living beings, including our maturing adolescents;

The emergence of abstractions and final mathematical formulae concerning the natural life force in the universe, and the right to their dissemination among one's fellow men;

The handling, use and distribution of instruments of basic research in any field, medical, educational, preventive, physical, biological, and in any fields which emerge from such basic activities and which, resting on such principles, must by all means remain free.

Attempts such as branding activities and instruments of such kind as "adulterated," in other words as fraud, only characterizes the narrowness of the horizon of the complainant.

No man-made law ever, no matter whether derived from the past or projected into a distant, unforeseeable future, can or should ever be empowered to claim that it is greater than the Natural Law from which it stems and to which it must inevitably return in the eternal rhythm of creation and decline of all things natural. This is valid, no matter whether we speak in terms such as "God," "Natural Law," "Cosmic Primordial Force," "Ether" or "Cosmic Orgone Energy."

The present critical state of international human affairs requires security and safety from nuisance interference with efforts toward full, honest, determined clarification of man's relationship to nature within and without himself; in other words, his relationship to the Law of Nature. It is not permissible, either morally, legally or factually to force a natural scientist to expose his scientific results and methods of basic research in court. This point is accentuated in a world crisis where biopathic men hold in their hands power over ruined, destitute multitudes.

To appear in court as a *"defendant"* in matters of basic natural research would in itself appear, to say the least, extraordinary. It would require disclosure of evidence in support of the position of the discovery of the Life Energy. Such disclosure, however, would invoke untold complications, and *possibly national disaster.*

Proof of this can be submitted at any time only to a duly authorized personality of the US Government in a high, responsible position.

Scientific matters cannot possibly ever be decided upon in court. They can only be clarified by prolonged, faithful bona fide observations in friendly exchange of opinion, never by litigation. The sole purpose of the complaint is to entangle orgonomic basic research in endless, costly legal procedures a la Panmunjom, which will accomplish exactly NOTHING rational or useful to human society.

Inquiry in the realm of Basic Natural Law is *outside the judicial domain,* of this, or ANY OTHER KIND OF SOCIAL ADMINISTRATION ANYWHERE ON THIS GLOBE, IN ANY LAND, NATION OR REGION.

Man's right to know, to learn, to inquire, to make bona fide errors, to investigate human emotions must, by all means, be safe, if the word FREEDOM should ever be more than an empty slogan.

If painstakingly elaborated and published scientific findings over a period of 30 years could not convince this administration, or will not be able to convince any other social administration* of the true

*Excepting, of course the "interest" displayed by the Soviet Union!—J. E.

nature of the discovery of the Life Energy, no litigation in any court anywhere will ever help to do so.

I, therefore, submit, in the name of truth and justice, that I shall not appear in court as the "defendant" against a plaintiff who by his mere complaint already has shown his ignorance in matters of natural science. I do so at the risk of being, by mistake,* fully enjoined in all my activities. Such an injunction would mean practically exactly nothing at all. My discovery of the Life Energy is today widely known nearly all over the globe, in hundreds of institutions, whether acclaimed or cursed. It can no longer be stopped by anyone, no matter what happens to me.

Orgone Energy Accumulators, the "devices" designed to concentrate the cosmic Orgone Energy, and thus to make it available to further research in medicine, biology and physics, are being built today in many lands, without my knowledge and consent, and even without any royalty payments.

On the basis of these considerations, I submit that the case against Orgonomy be taken out of court completely.

WILHELM REICH, M.D.
Chairman of Basic Research
of THE WILHELM REICH FOUNDATION

*In my opinion, the basic "mistake" of the court was its failure to examine the voluminous literature on Orgonomy which easily disproves the allegations of fraud brought by agents of the FDA. This literature was freely available to the court. Once the judge's signature was summarily placed upon the injunction, it set into motion the events that led to the strangulation of basic natural research in the United States, and the imprisonment and death of Wilhelm Reich.—J. E.

Appendix B

IN THE UNITED STATES DISTRICT COURT
FOR THE DISTRICT OF MAINE
SOUTHERN DIVISION

UNITED STATES OF AMERICA

Plaintiff

v.

THE WILHELM REICH FOUNDATION, a Maine Corporation, WILHELM REICH and ILSE OLLENDORFF,

Defendants

Civil Action No. 1056

DECREE OF INJUNCTION

Plaintiff having filed a Complaint for Injunction herein to enjoin the defendants and others from further alleged violations of the Federal Food, Drug, and Cosmetic Act; and each defendant having been duly served, on February 10, 1954, with a summons and copy of the Complaint; and no defendant having appeared or answered in person or by representative, although the time therefor has expired; and each defendant having been duly served, on February 26, 1954, with a copy of Requests for Admissions; and no defendant having served any answer to said Requests, although the time therefor has expired; and the default of each defendant having been entered therein; and it appearing that the defendants, unless enjoined therefrom, will continue to introduce or cause to be introduced or delivered, or cause to be delivered into interstate commerce orgone energy accumulators, devices within the meaning of the Federal Food, Drug, and Cosmetic Act, 21 U.S.C. 331 (a) and (k); and the Court having been fully advised in the premises:

IT IS HEREBY ORDERED, ADJUDGED, AND DECREED that the defendants, THE WILHELM REICH FOUNDATION, WILHELM REICH, and ILSE OLLENDORFF and each and all of their officers, agents, servants, employees, attorneys, all corporations, associations, and organizations, and all persons in active concert or participation with them or any of them, be, and they hereby are, perpetually enjoined and restrained from doing any of the following acts, directly or indirectly, in violation of Sections 301(a) or 301(k) of the Federal Food, Drug, and Cosmetic Act (21 U.S.C.C. 331(a) or (k)) with respect to any orgone energy accumulator device, in any style or model, any and all accessories, components or parts therefrom, or any similar device, in any style or model, and any device purported or represented to collect and accumulate the alleged orgone energy:

(1) Introducing or causing to be introduced or delivering or causing to be delivered for introduction into interstate commerce any such article of device which is:

(a) Misbranded within the meaning of Section 502(a) of the Act (21 U.S.C. 352(a)) by reason of any representation or suggestion in its labeling which conveys the impression that such article, in any style or model, is an outstanding therapeutic agent, is a preventive of and beneficial for use in any disease or disease condition, is effective in the cure, mitigation, treatment, and prevention of any disease, symptom, or condition; or

(b) Misbranded within the meaning of Section 502(2) of the Act (21 U.S.C. 352(a)) by reason of any representation of suggestion in its labeling which conveys the impression that the alleged orgone energy exists; or

(c) Misbranded within the meaning of Section 502(a) of the Act (21 U.S.C. 352 (a)) by reason of any photographic representation or suggestion with a caption, or otherwise, which conveys the impression that such is an actual photograph depicting the alleged orgone energy or an alleged excited orgone energy field; or

(d) Misbranded within the meaning of Section 502(a) of the Act (21 U.S.C. 352(a)) by reason of any other false or misleading representation or suggestion; or

(e) Adulterated within the meaning of Section 501(c) of the Act (21 U.S.C. 351(c)) in that (1) its strength differs from or its quality falls below that which it purports or is represented to possess or (2) it purports to collect from the atmosphere and accumulate in said device the alleged orgone energy; or

(2) Doing any act or causing any act to be done with respect to any orgone energy accumulator device while such device is held for sale (including rental, or any other disposition) after shipment in interstate commerce which results in said device becoming misbranded or adulterated in any respect; and

IT IS FURTHER ORDERED:

(1) That all orgone energy accumulator devices, and their labeling, which were shipped in interstate commerce and which (a) are on a rental basis, or (b) otherwise owned or controlled by any one of the defendants, or by the defendants, be recalled by the defendants to their place of business at Rangeley, Maine; and

(2) That the devices referred to in (1) immediately above, and their parts, be destroyed by the defendants or, they may be dis-

mantled and the materials from which they were made salvaged after dismantling; and

(3) That the labeling referred to in paragraph (1), just above, except those items for which a specific purchase price was paid by their owners, be destroyed by the defendants; and

(4) That all parts or portions of orgone accumulator devices shipped in interstate commerce and returned to Rangeley, Maine, or elsewhere, and awaiting repair or re-shipment be destroyed by the defendants, or they may be dismantled and the materials from which they were made salvaged after dismantling; and

(5) That all copies of the following items of written, printed, or graphic matter, and their covers, if any, which items have constituted labeling of the article of device, and which contain statements and representations pertaining to the existence of orgone energy accumulators, and the use of such alleged orgone energy by employing said accumulators in the cure, mitigation, treatment, and prevention of disease, symptoms and conditions:

The Discovery of the Orgone by Wilhelm Reich
Vol. I—The Function of the Orgasm
Vol. II—The Cancer Biopathy
The Sexual Revolution by Wilhelm Reich
Ether, God and Devil by Wilhelm Reich
Cosmic Superimposition by Wilhelm Reich
Listen, Little Man by Wilhelm Reich
The Mass Psychology of Fascism by Wilhelm Reich
Character Analysis by Wilhelm Reich
The Murder of Christ by Wilhelm Reich
People in Trouble by Wilhelm Reich

shall be withheld by the defendants and not again employed as labeling; in the event, however, such statements and representations, and any other allied material, are deleted, such publications may be used by the defendants;* and

(6) That all written, printed, and graphic matter containing instructions for the use of any orgone energy accumulator device, instructions for the assembly thereof, all printed and other announcements and order blanks for the items listed in the paragraph immediately above, all documents, bulletins, pamphlets, journals, and booklets entitled in part, as follows: CATALOGUE SHEET, PHYSICIAN'S REPORT, APPLICATION FOR THE USE OF THE ORGONE ENERGY ACCUMULATOR, ADDITIONAL INFORMATION REGARDING SOFT ORGONE IRRADIATION, ORGONE ENERGY ACCUMULATOR ITS SCIENTIFIC AND MEDICAL

*Despite the injunction, some of this enjoined literature has been reprinted in its original form, without deletions, by Farrar Straus and Giroux.—J. E.

USE, ORGONE ENERGY BULLETIN, ORGONE ENERGY EMERGENCY BULLETIN, INTERNATIONAL JOURNAL OF SEX-ECONOMY AND ORGONE RESEARCH, INTERNATIONAL ZEITSCHRIFT FUR ORGONOMIE, EMOTIONAL PLAGUE VERSUS ORGONE BIOPHYSICS, ANNALS OF THE ORGONE INSTITUTE, and ORANUR EXPERIMENT, but not limited to those enumerated* shall be destroyed; and

(7) That the directives and provisions contained in paragraphs (1) to (6) inclusive, above, shall be performed under the supervision of employees of the Food and Drug Administration, authorized representatives of the Secretary of Health, Education and Welfare; and

(8) That for the purpose of supervision and securing compliance with this decree the defendants shall permit said employees of the Food and Drug Administration, at reasonable times, *to have access to and copy from, all books, ledgers, accounts, correspondence, memoranda, and other records and documents* [italics mine—J.E.] in the possession or under the control of said defendants, including all affiliated persons, corporations, associations, and organizations, at Rangeley, Maine, or elsewhere, relating to any matters contained in this decree. Any such authorized representative of the Secretary shall be permitted to interview officers or employees of any defendant, or any affiliate, regarding any such matter subject to the reasonable convenience of any of said officers and employees of said defendants, or affiliates, but without restraint or interference from any one of said defendants; and

(9) That the defendants refrain from, either directly or indirectly, in violation of said Act, disseminating information to the assembly, construction, or composition of orgone energy accumulator devices to be employed for therapeutic or prophylactic uses by man or for other animals.

March 19, 1954
2:45 P.M.

/s/ JOHN D. CLIFFORD, JR.
United States District Judge
for the District of Maine.

A true copy of original filed at 2:45 P.M. on March 19, 1954.
ATTEST:
/s/ Morris Cox
Clerk, United States District Court

*Why wasn't *The Einstein Affair* listed? No mention, no attention?—J. E.

Appendix C

Physicians, scientists, and educators who corroborated the work of Wilhelm Reich and whose articles appeared in the literature impounded and burned by the FDA injunction:

Anderson, William A., M.D., surgeon, Bloomfield, N.J.
Atkin, R. H., mathematical physicist, London, England.
Baker, Elsworth F., M.D., formerly Chief of Women's Service, New Jersey State Hosp., Marlboro, N.J.
Baumann, Jacob, psychologist, Zurich, Switzerland.
Bremer, Kenneth M., M.D., Orange Memorial Hosp., N.J.
Cott, A. Allan, M.D., formerly Assistant Chief of Women's Services, New Jersey State Hosp., Marlboro, N.J.
Gold, Philip, M.D., formerly Psychiatrist, Mental Hygiene Clinic, Bellevue Hosp., N.Y.C.
Hamilton, A. E., Director, Hamilton School, Sheffield, Mass.
Hoppe, Walter, M.D., physician and Chief Psychiatrist, Tel Aviv (formerly with Psychiatric Clinic, Bonhoeffer, Berlin).
Levine, Emanuel, M.D., formerly Staff Psychiatrist, Vermont State Hosp. for the Insane, Waterbury, Vt.
McDonald, Helen E., Ph.D., formerly Head of Science Department, William Woods College, Fulton, Mo.; Teaching & Research Fellow, Univ. of Calif., Berkeley, Calif.
Neill, A.S., Director, Summerhill School, Leiston, Suffolk, England.
Oller, Charles I., M.D., Diplomate of the American Board of Neurology and Psychiatry; Associate in Neurology, Graduate School of Medicine, Univ. of Pennsylvania.
Raknes, Ola, Ph.D., psychologist; Lecturer, 13th International Congress of Psychology, Stockholm, Sweden.
Raphael, Chester M., M.D., formerly Senior Resident Psychiatrist, New Jersey State Hosp., Marlboro, N.J.
Sharaf, Myron, M.A., Teaching Fellow in Education and Research Assistant in Psychology, Harvard Univ.
Singer, Richard, M.D., formerly with the Jersey City Medical Center; Bellevue Psychiatric Hosp. and Hillside (N.J.) Hosp.
Silvert, Michael, M.D., Psychiatrist, V.A. Administration Mental Hygiene Clinic, N.Y.C.; former Resident in Psychiatry, Menninger Foundation School of Psychiatry, Topeka, Kan.
Sobey, Victor M., M.D., formerly Staff Psychiatrist, Newark Mental Hygiene Clinic, Newark, N.J., and Union County Mental Hygiene Clinic, Union County, N.J.
Weverick, N., M.D., Montreal, Canada.

Willie, James A., M.D., formerly Senior Psychiatrist, New Hampshire State Hosp., Concord, N.H.

Wolfe, Theodore P., M.D., Associate, Columbia University (N.Y.C.), and Psychiatrist and Research Physician, Presbyterian Hosp., N.Y.C.

Appendix D

Following are excerpts from the brief submitted by Wilhelm Reich to the Court of Appeals for the First Circuit. A complete summation of the entire case can be found in the excellent article "United States of America v. Wilhelm Reich," by David Blasband, A.B., LL.B., which appears in vol. 1, and vol. 2, no. 1, of the *Journal of Orgonomy,* 1967 and 1968 (see Bibliography).

PART I–APPEAL TO FAIRNESS

A. General Technical Grounds of Appeal and Argument

The judgement against the Discoverer of the Life Energy of May 7th, 1956, was obtained under legal conditions and social circumstances which are without precedent in the history of jurisprudence. These extraordinary conditions and circumstances blocked the defendants *technically* in clarifying fully the issue before the jury. The jury rendered its verdict uninformed. Had the full factual evidence involved in the case been presented, the verdict would have been *"not guilty."* However, this evidence was submerged in and blocked by *top secret* involvements of the basic research work of *Wilhelm Reich, M.D., the Discoverer of the primordial, massfree cosmic energy* (also Life Energy). . . .

B. General Purpose of Appeal

This is an appeal to rescind judgment in an *extraordinary* case, *without precedent* in the annals of jurisprudence, unprecedented in Fact and in Law. Never before has mankind faced a situation such as underlies this case. On November 4th, 1955, information appeared in "C.R.I.F.O. Orbit":

The Case for Interplanetary 'War'

Space War Possible is MacArthur Hint

"Supporting CRIFO's beliefs are the timely and sobering words voiced by *General Douglas MacArthur,* before visiting Mayor of Naples, Achille Lauro on October 7, 1955 at the Waldorf-Astoria, New York. The Mayor revealed the General's statements to the New York Times as follows: '. . . *He thinks that another war would be double suicide* and that there is enough sense on both sides of the Iron Curtain to avoid it. . . . *He believes that because of the development of science all countries on earth will have to unite to survive and to make a common front against attack by people from* other planets. . . .' The Mayor added that in the General's opinion the politics of the future will be cosmic or interplanetary."

Substantial matters of a top secret nature, without precedent in fact and law were involved in the legal conflict before the court:

(A) The Oranur Experiment (Begun January 5, 1951, still running)

Never before has the atmospheric energy of this planet been irritated by nuclear energy as in Oranur: Minimal amounts of NR* irritate great concentrations of OR† energy. (In the atomic explosion, tremendous amounts of NR energy irritate unconcentrated atmospheric energy, unknown to classical science except as "Static.") . . .

(B) The Problem of Invaders from Outer Space

Never before in the scientific development of mankind has the intrusion of visitors from outer space into the earth's atmosphere posed such basically unformulated problems for natural science. Never before has routine technology been as unprepared as in 1951 to cope with such problems. And never before has a case been tried by a jury with the exclusion of such facts which determined the conduct of the defendants all through the injunction.

(C) The Integration of Oranur and Space Problem

Never before in the history of man's development have two such revolutionary, independent events merged into one perfect integration. The Discoverer of the Cosmic Energy was until 1953 totally unaware of the grave bearings his research had on the invasion from outer space.

Never before has human society been as unprepared to meet such grave problems: never have the peoples of the earth been unprepared emotionally to accept the challenge of both the invasion from outer space and the corresponding discovery of the Cosmic Energy used in those space ships. The critical economic, social, political, legislative problems, too, involved here are without precedent in the history of man. . . .

2. Legal Situation Without Precedent

May I illustrate the unprecedented legal circumstances with known ones.

We suppose that no one has heard anything about a *Cosmic*

---

*NR is nuclear energy.—J. E.

†OR is Orgone energy. Reich used only *one milligram of radium* to produce the Oranur effect. Barry Commoner in *Science and Survival* (Ballantine Books, New York, 1966) estimates that between 1948 and 1962, planetary nuclear tests "introduced into the environment radioactivity [from strontium-90 *only*] equivalent to about *ONE BILLION GRAMS OF RADIUM!*"—J. E. Italics added.

Energy Motor, as it will be used in space travel; but everyone knows the mechanical principle of *"thrust"* as used in jet propulsion.

Let us return to the time years ago when no one, except those working on jet thrust engines, knew anything about this type of motion, then new, too.

Let us further imagine that an administrative agency of the government, organized originally to administer steam engines, were, under pressure of the steam engine industry, to declare that thrust is a fraud, that *"it does not exist."* Imagine furthermore that this steam engine administration would put in a complaint in court against the thrust engineers, claiming "fraud." The court has never heard of the principle of *"thrust."* The court would therefore know nothing about it. There are no *statutory* laws on the books to tell about the legal situation of thrust engines as yet embryonic, untested knowledge. The complainant, the steam engine administrator, has succeeded in skillfully using slander that has been spread about the thrust engineers by the competing steam engine manufacturers. . . .

In addition, thrust engine research has been classified by another government agency as top secret, a fact which is probably known to the complainant, but he fails to communicate it to the court.

The case of the *noiseless,* Cosmic Energy Motor Engines is in exactly the same position as presented here with respect to thrust engines. . . . However, there is one important difference: The realm of the future cosmic technology is so much vaster than that of mechanical thrust in the earth's atmosphere. . . .

The question "Why" the engineers did not appear in court is crucial and pertinent to the special case. The injunction did not concern a routine case of fraudulent production to deceive the public. It was, on the contrary, the most crucial discovery ever made in natural science by an acknowledged, widely-known scientist and physician, arbitrarily misrepresented to the court as a quack and fraudulent crook. . . .

It is characteristic of the unprecedented nature of the case that the acting U.S. attorney, Peter Mills, has failed, before issuing and signing the complaint for injunction, to check the complaint with regard to the exactness of the quotations from the literature to be banned and the claims allegedly made for cure. This omission emerges clearly from the records of the trial evidence: The books impounded as *"labeling"* of a *"fraudulent"* device were either published long before the actual discovery: "Character Analysis" (1928-1937), "Sexual Revolution" (1930-1935), "Mass Psychology of Fascism" (1933), or were dealing with matters unrelated to medicine, such as "Cosmic Superimposition," "Ether, God and Devil," "Listen, Little Man," "CORE." . . .

What interest may U.S. Government officials have to mislead U.S. Courts so cunningly, so persistently, and so successfully so far?

Now, the U.S. Air Force has full knowledge of the invasion of the earth's atmosphere, of invaders using unheard-of kinds of motion; the AAF also became aware of the discovery of a new motor force on this earth by Wilhelm Reich. The AAF undertook to study and to protect the discovery of the primordial cosmic force. Without this new force there seemed no hope to cope with the invader from outer space in an adequate technological manner. However, the discovery of the Cosmic Energy within and without the organism constitutes a revolutionary event in the history of the human race; havoc and chaos would occur should this discovery not be handled like a highly explosive bomb, more dangerous than the H-bomb. . . .

Political interests of a foreign power in combination with national U.S.A. commercial interests entered the scene of orgonomic research when, in 1948, the *orgonomic motor function* was discovered in my laboratory and documents regarding this discovery were executed in the law office of *Mills and Mills, including Peter Mills,* the later prosecutor of the Discovery. This OR-Motor (first document was notarized by Peter Mills) was the first step in the technological development toward noiseless, smokeless, smooth-functioning locomotor machines of the future, the Cosmic Energy Motor. Inevitably, this new kind of motor will replace the steam engine and the electric motor. It will be fed by the practically limitless power resources contained in the Cosmic Energy Ocean. It will be the motor that will carry our Space Ships into vastnesses as yet unimaginable. . . .

No amount of proof has been enough in eight years of struggle around the UFO problem to convince the Enemy of Man in mechanistic science of a clear-cut reality. No amount of proof will ever convince the mechanistic heir of Stalinite methods. The work on our space problems is stuck in a mire of *evasion* of the essential. Backslapping and fake socialities, "making no enemies," avoiding ridicule at all costs seems more important than the existence of the human race.

"What Constitutes Proof?" The answer is: *Nothing,* as long as the Enemy of Man is permitted to foul up everything important on the social or scientific scene.

"What do you want?" one officer, hardworking and harassed by the pressure from both outer space and a few frightened Little Men asked: *"Does a UFO have to come in and land on your desk at the ATIC?"*

Nothing will convince them. They are frightened. Orgonomy would have long since gone under had it not resolved to let the "authorities" come to the pioneers in cosmic research *to learn,* before they could judge. The public will get used to these new ways. Before Humanity can hope to cope with the space problem it will needs have to *reconstruct itself* and society.

However, to judge from the experience of more than a dozen workers in orgonomy during the five years of the DOR emergency, man on earth will needs grow up to his fate; moreover, he will not escape the power of Oranur which has gripped our planet. *The Biological Revolution Of Man Is On.*

Bibliography

Books by Wilhelm Reich, currently available

Character Analysis, 3rd ed. New York: Farrar Straus and Giroux.
Function of the Orgasm, 2nd ed. New York: Farrar Straus and Giroux.
Listen, Little Man! New York: Farrar Straus and Giroux.
The Murder of Christ: The Emotional Plague of Mankind. New York: Farrar Straus and Giroux.
Selected Writings. New York: Farrar Straus and Giroux.
The Sexual Revolution. New York: Farrar Straus and Giroux.

By Wilhelm Reich, banned or burned

The Cancer Biopathy New York: Orgone Institute Press, 1948.
CORE (Cosmic Orgone Engineering), "OROP DESERT: Part 1. Space Ships, DOR and Drought" (1954). Orgonon, Rangeley, Maine: Orgone Institute Press.
CORE, vol. 7, nos. 1-2, March 1955. Includes published findings of physicians and scientists other than Reich.
Cosmic Superimposition. Orgonon, Maine: Wilhelm Reich Foundation, 1951.
The Einstein Affair (Documentary Volume A-XI-E). Orgonon, Maine: Orgone Institute Press, 1953.
Ether, God and Devil. Orgonon, Maine: Orgone Institute Press, 1949.
The Mass Psychology of Fascism. New York: Orgone Institute Press, 1946.
The Oranur Experiment: First Report (1947-51). Orgonon, Maine: Wilhelm Reich Foundation, 1951.
People in Trouble. Orgonon, Maine: Orgone Institute Press, 1953.

Court Proceedings

Supreme Court of the United States, October Term, 1956: No. 688, "Petition for a Writ of Certiorari to the U.S. Court of Appeals for the First Circuit," Washington, D.C., Jan. 10, 1957.
Supreme Court of the United States, October Term, 1956: No. 688, "Petitioner's Reply Brief," Washington, D.C. Feb. 18, 1957.

Based on Orgonomy and Wilhelm Reich

Baker, Elsworth F., M.D., *Man in the Trap.* New York: Macmillan, 1967.
Bean, Orson, *Me and the Orgone.* New York: St. Martin's Press, 1971.

Eden, Jerome, *Suffer the Children*. Mt. Vernon, N. Y.: Eden Press, 1959.

Journal of Orgonomy (a scientific journal published since 1967), Orgonomic Publications, Inc., Box 476, Ansonia Station, New York, N. Y. 10023.

Raknes, Ola, *Wilhelm Reich and Orgonomy*. New York: St. Martin's Press, 1970.

Reich, Ilse Ollendorff, *Wilhelm Reich: A Personal Biography*. New York: St. Martin's Press, 1969.

Reich Speaks of Freud, ed. Mary Higgins and Chester M. Raphael, M.D. New York: Farrar Straus and Giroux, 1967.

Wolfe, Theodore P., *The Emotional Plague Versus Orgone Biophysics*. New York: Orgone Institute Press, 1948. (Destroyed.)

On Nuclear Radiation and Atomic Hazards

Curtis, Richard, and Hogan, Elizabeth, *Perils of the Peaceful Atom*. New York: Ballantine Books, 1969.

Gofman, John W., M.D., and Tamplin, Arthur R., Ph.D., *Poisoned Power*. Emmaus, Pa.: Rodale Press Inc., 1971.

Russell, Walter and Lao, *Atomic Suicide*. Swannanoa, Waynesboro, Va.: Univ. of Science and Philosophy, 1957.

Schubert, Jack, and Lapp, Ralph, *Radiation*. New York: Viking Press, 1957.

Snyder, Ernest E., Ed.D., *Please Stop Killing Me!* New York: New American Library, 1971.

On Soviet Nuclear Espionage

Toledano, Ralph de, *The Greatest Plot in History*. New York: Duell, Sloan and Pearce. Reprinted in *Conservative Book Club Omnibus*, vol. 2, New Rochelle, N. Y.

On Unidentified Flying Objects

Binder, Otto O., *Flying Saucers Are Watching Us*. New York: Belmont Productions, 1968.

———, *What We Really Know About Flying Saucers*. Greenwich, Conn.: Fawcett, 1967.

Chambers, Howard V., *The Facts on the Flying Saucer Controversy*. New York: Grosset and Dunlap, 1968.

Edwards, Frank, *Flying Saucers—Here and Now*. New York: Bantam Books, 1967.

———, *Flying Saucers—Serious Business*. New York: Bantam Books, 1967.

Fuller, John G., *The Interrupted Journey*. New York: Dell, 1967.

Lorenzen, Coral E., *Flying Saucers: The Startling Evidence of the Invasion from Outer Space*. New York: New American Library, 1966.

Lorenzen, Coral and Jim, *UFOs: The Whole Story*. New York: New American Library, 1969.

Lorenzen, Jim and Coral, *UFOs Over the Americas*. New York: New American Library, 1968.

Steiger, Brad, and Whritenour, Joan, *Flying Saucers Are Hostile*. New York: Universal Publ. and Distrib. Corp., 1967.

Vallee, Jacques and Janine, *Flying Saucers a Challenge to Science*. Chicago: Regnery, 1966.

Wilkins, Harold T., *Flying Saucers Uncensored*. New York: Pyramid Books, 1967.